# The Ing Engine *of* Reality

# The Ingenious Engine of Reality

A *Trousers Of Reality* Book

**Challenge your habits and free yourself
by discovering the lessons of neuroscience
to understand yourself and other people better.**

### Barry Evans
MBCS CEng CITP

Code Green Publishing
Coventry, UK

Copyright © 2011 by Barry Evans

Barry Evans asserts the moral right to be identified as the author of this work.

A catalogue record for this book is available from the British Library.

All rights reserved; no part of this publication may be reproduced, stored in a retrieval system, transmitted, in any form, or by any means electronic, mechanical, photocopying, recording or otherwise, without the prior written permission of the Publisher, nor be circulated in any form of binding or cover other than that in which it is published and without a similar condition including this condition being imposed on the subsequent purchaser.

ISBN 978-1-907215-19-3

Previously published as Managing Knowledge

All illustrations by the Author unless otherwise stated
Cover design by the Author

Version 1.2

Published by Code Green Publishing
www.codegreenpublishing.com

# The Ingenious Engine of Reality

In Greek mythology Sisyphus is the mortal who outwits Zeus, Thanatos, Hades and Aries (Authority, Death, Hell and War). His reward in the afterlife is to roll a rock to the summit of a hill only to have it roll down again and to repeat forever in an endless, thankless, pointless task.

In a footnote to the legend the French philosopher Albert Camus, in his essay: "The Myth of Sisyphus", compares lives spent at work in the modern world to the futile and meaningless labours of Sisyphus. Camus concludes that only when Sisyphus realises the absurd nature of his predicament is he freed to reach a state of acceptance.

Life can be pretty absurd. It is recursive and often contains unavoidable repetition and futility. Like Sisyphus we can continue to roll our boulder out of meaningless habit. We can wear a rut or we can find intrinsic value in the task and learn something from each iteration. We can change our reality by degrees.

Like Sisyphus we can find freedom of mind by learning from things that seem to imprison us in circumstance. I like to think that Sisyphus represents that in us which dares to challenge the status quo and make a break for freedom in the face of mortality – he defied death. We defy our limitations with the tools of science, knowledge and enlightenment.

We are ingenious machines well suited to our task by evolution. Our brains react to input and shape themselves on the basis of our reaction. Our genes react to our environment and behaviour to shape our physical selves and abilities. What you are, how you think and how well you will live, are greatly dependent on the quality of your thinking. The quality of your thinking is greatly dependent on what you have convinced yourself to believe and what you value. Your values and beliefs ignite the engine of reality in your head.

Your brain is the engine of reality. Although it is always in your head and surging with lightning every moment of your life you need to switch it on and turn it up to full power so that it becomes a dynamo of creativity. When it is on it works for us by managing how we react to what happens to us. It decides what we learn, how we use what we know and who we become.

# Table Of Contents

SECTION ZERO: OVERTURES ........................................................................... 1
- **The Music Of Change** .................................................................................. 3
- **Contextualisation** ....................................................................................... 6
- **How To Read This Book** ............................................................................. 8
- **A Nested Koan** .......................................................................................... 11
  - *Let Information Flow Through Chaos* ........................................................ 11
  - *Integrate Feedback With Connections* ....................................................... 11
  - *Find Choice Where There Are Intersections* .............................................. 11
  - *Examine And Refine The Loop* ................................................................ 11

SECTION ONE: PERPETUALLY BECOMING ................................................. 13
- **First Calibration** ....................................................................................... 15
  - *It Is Through Doing That We Become* ...................................................... 15
- **Neuroprogramming** .................................................................................. 19
  - *Discovering Neuroscience* ....................................................................... 19
  - *Neuroplasticity* ........................................................................................ 20
  - *The Important Principles Of Neuroplasticity* ............................................. 21
  - *Doing And Thinking* ................................................................................ 23
  - *Determinism Is A Choice* ........................................................................ 23
  - *Free Will Is A Habit* ................................................................................ 24
  - *Essential Things To Know About Your Brain* .......................................... 26
  - *Summary* ................................................................................................ 31
  - *Conclusion* .............................................................................................. 31
- **Mapping The Little Grey Cells** .................................................................. 32
  - *Neuro Progress* ........................................................................................ 32
  - *The Brain Is A Feedback Mechanism* ...................................................... 46
- **The Storm Of Ideas** ................................................................................... 52
  - *The Things Being Talked About* .............................................................. 52
  - *The Brain Storm* ...................................................................................... 52
- **Shaping Reality** ......................................................................................... 54
  - *The Resilience Of The Brain* .................................................................... 54
  - *Summary* ................................................................................................ 56
  - *Conclusion* .............................................................................................. 56

SECTION TWO: THE NATURE OF KNOWLEDGE ........................................ 57
- **Second Calibration** ................................................................................... 59
  - *The Spice Of Life* .................................................................................... 59
- **Managing Philosophy** ............................................................................... 61
  - *The Dragons Tail* ..................................................................................... 61
  - *Stalking The Dragon* ............................................................................... 61
  - *A Flight Toward Reason* .......................................................................... 62
- **Empathise With Yourself** .......................................................................... 64
  - *Control Groups* ........................................................................................ 64
  - *Human Difference Engines* ...................................................................... 65
  - *Facing Our Fears* ..................................................................................... 66
  - *Embedded Perceptions* ............................................................................ 67
  - *The Russian Doll Self* .............................................................................. 68

 *Reinforcing Mistakes*...................................................................................*69*
 *Practising Perfection*...................................................................................*70*
 *Empathising With Others*............................................................................*71*
**The Nature Of Consciousness**...........................................................**73**
 *Interdependence Of Thought*.......................................................................*73*
 *The Internal Judiciary*..................................................................................*74*
 *The Executive Mind*.....................................................................................*74*
 *Learn To Relax And Love The Unconscious*................................................*75*
**Arm Yourself With A Priori Knowledge**..........................................**77**
 *Knowledge Is A Powerful Tool*....................................................................*77*
 *Philosophy Unbound*...................................................................................*78*
 *Complex Equivalence*..................................................................................*79*
 *Reality As A Metaphor*................................................................................*79*
**The Crucible**........................................................................................**81**
 *Suffering*.......................................................................................................*81*
 *The Anvil Of Life*.........................................................................................*82*
 *Suffering Dislodges Certainties*...................................................................*82*
 *Learning Without Suffering*.........................................................................*83*
 *Transforming Suffering Into Effortless Gain*..............................................*84*
 *Clarifying Equivalence And Re-Associating*...............................................*85*
 *Dissociation And Models*............................................................................*87*
 *Professional Detachment*............................................................................*88*
 *Objective Associations*................................................................................*89*
 *Avoiding The Crucible*................................................................................*89*
 *Delayed Gratification And Selfishness*.......................................................*90*
 *Intrinsic Value*.............................................................................................*92*
 *Money Is Not A Motivator, It Is A Demotivator*........................................*93*
 *Balance, Counterbalance And Pay*.............................................................*94*
**The Reservoir**.....................................................................................**98**
 *Connecting To Experience And Knowledge*................................................*98*
 *The Quality Of Knowledge Is Filtered*......................................................*101*
**The Shape Of Knowledge**................................................................**103**
 *Mathematics, Logic And Philosophy*........................................................*103*
 *Infinite Paradox Machine*..........................................................................*104*
 *Russell's Paradox Revisited*.......................................................................*107*
 *Energy Horses*............................................................................................*108*
 *On Newton's Beach*....................................................................................*109*
 *Deal Or No Deal*........................................................................................*110*
 *Getting Involved*.........................................................................................*111*
 *Conclusion*.................................................................................................*113*
 *Summary:*...................................................................................................*114*
**Managing Knowledge**......................................................................**116**
 *The Information Age Effect*......................................................................*116*
**Three Simple Ideas**..........................................................................**120**
**Evolution Of Knowledge I**...............................................................**121**
**Evolution Of Knowledge II**.............................................................**122**
**Evolution Of Knowledge III**............................................................**123**
**The Root Of Knowledge**..................................................................**124**
 *Thoughts*.....................................................................................................*124*
 *The Paradox Of Thought*...........................................................................*125*

*Reality Is A Magician*.................................................................................................*127*
*The Rational Explanation*.........................................................................................*128*
**Filtering**...............................................................................................................**130**
   *Context*.......................................................................................................*130*
   *Filters*.........................................................................................................*130*
   *Mach's Conjecture*.....................................................................................*130*
   *Boundaries As Context*..............................................................................*131*
   *Atomic Distinctions*...................................................................................*132*
   *Knowledge Is A Product Of Evolution*.......................................................*132*
   *Filtering And Curious Water*.....................................................................*133*
   *The Two Faces Of Filtering*.......................................................................*136*
**Questions Are Filters**...........................................................................................**138**
   *Framing The Problem*................................................................................*138*
   *Process Grows*...........................................................................................*138*
   *Programming Precedent*............................................................................*139*
   *The Programmed Mind*..............................................................................*139*
   *Living In Free-Fall*....................................................................................*140*
   *The Human Advantage*..............................................................................*141*
   *The Two Faces Of Versatility*....................................................................*142*
   *Working Models Are Our Speciality*.........................................................*143*
   *Asking For What You Want*......................................................................*143*
   *Filtering Tools*...........................................................................................*144*
**Intersecting**..........................................................................................................**146**
   *Balance*......................................................................................................*146*
   *Depth Of Perception*..................................................................................*146*
   *Triangulation*.............................................................................................*147*
   *Triangulations And The Power Of 3*.........................................................*148*
**Connecting**..........................................................................................................**150**
   *Inspiration*.................................................................................................*150*
   *Peak Experience*........................................................................................*150*
   *Connection Is The Mechanics Of Inspiration*...........................................*151*
   *Breaking Eggs To Make Omelettes*...........................................................*153*
   *How To Connect*.......................................................................................*154*
   *Einstein's Brain*.........................................................................................*156*
   *The Prepared Mind*...................................................................................*157*
**The Paradox Of Growth**.....................................................................................**159**
   *The Uncertainty Paradox*..........................................................................*159*
   *Pruning Knowledge*...................................................................................*159*
   *Example*.....................................................................................................*160*
   *The Ladder Of Life*...................................................................................*161*
   *Balance Is A Process Not A State*.............................................................*162*
   *Have You Forgotten Project Management?*..............................................*162*
**The Phoenix Of Experience**................................................................................**163**
   *Personal Invention*....................................................................................*163*
   *Personal Reinvention*................................................................................*163*
   *Successful Projects Iterate*........................................................................*164*
   *Perception Is Projection*...........................................................................*165*
   *Update The Projector*...............................................................................*167*
   *Incremental Cycles Breed Feedback And Chaos*.....................................*168*
SECTION THREE: THE ART OF KNOWLEDGE.................................................................173

**Third Calibration** ........................................................................................ **175**
   *Models – The Basis Of Consciousness* ......................................................... *175*
   *Q&A* ................................................................................................................ *178*
**Useful Concepts** ........................................................................................ **182**
   *Introduction* .................................................................................................. *182*
   *Stealing Fire From The Gods* ....................................................................... *182*
   *Secondary Gain* ............................................................................................ *182*
   *Epigenetics* .................................................................................................... *184*
   *Mirror Neurons* ............................................................................................. *187*
   *Theory Of Mind* ............................................................................................ *190*
   *Boundaries And Constraints* ........................................................................ *192*
   *Response Ability* ........................................................................................... *193*
   *A Priori Tea* .................................................................................................. *195*
   *Nominalisation And Denominalisation* ........................................................ *199*
   *Strange Loops* ............................................................................................... *200*
**Envisioning Creative Thinking** ................................................................ **205**
   *As If Slicing Reality* ..................................................................................... *205*
   *The Slices Out Of Context* ............................................................................ *207*
   *Logical Wormholes* ....................................................................................... *208*
   *Vertical And Lateral Thinking* ..................................................................... *210*
   *Access Genius – Reconnect The Slices Of Reality* ...................................... *212*
   *Creative Thinking – A Conclusion* ............................................................... *214*
**Feedback Loops** ....................................................................................... **215**
   *Learning Through Feedback* ........................................................................ *215*
   *The Cause – Effect Ecology* ......................................................................... *218*
   *Recognising Interference* .............................................................................. *225*
   *Self Defence* .................................................................................................. *226*
   *Out Of Complexity* ........................................................................................ *227*
**Order And Chaos** ..................................................................................... **228**
   *Butterflies And Cats* ..................................................................................... *228*
   *Deterministic Universes* ............................................................................... *228*
   *Deterministic Chaos Theory* ........................................................................ *229*
   *Sensitivity To Initial Conditions* ................................................................... *229*
   *A Simple Complexity* .................................................................................... *230*
   *Fractals* ......................................................................................................... *231*
   *Self Similarity* ............................................................................................... *231*
   *Recursion* ...................................................................................................... *231*
   *Chaotic Learning* ......................................................................................... *232*
   *Management Chaos* ..................................................................................... *232*
   *Consequences Of Chaos* .............................................................................. *237*
   *Complex Problems Usually Have Simple Solutions* .................................... *238*
   *Command And Control Is A Complex Solution To A Complex Problem*
   *That Requires A Simple Solution* ................................................................. *238*
   *Heuristic And Algorithmic Work* .................................................................. *239*
   *Processes Can Get Stuck In A Fight Or Flight State* .................................. *240*
   *NLP As A Chaos Utilisation Tool* ................................................................ *240*
   *Attributes Of Reality Viewed Through Chaos* ............................................. *241*
**Thinking Inside The Box – The Treehouse Of Truth** ............................... **242**
   *Boundaries, Counterbalance And Change* .................................................. *242*
   *Creating Space For Stability* ....................................................................... *243*

 *Willing Belief Is A Tool*...........243
 *Make Sure The Tree-House Has A Ladder: Reinstating Disbelief*...........244
 *These Beliefs And Behaviours Are Strategies*...........245
 *Learning Is A Recursive Process*...........245
 *Phobias Metaphor To Highlight The Importance Of 360 Degree Learning*...........246
 *The Principle Of Counterbalance*...........247
 *Process And Feedback*...........253
 *Feedback Engineering*...........255
 *Scaling Process*...........258
 *Productivity And The Rights Of The Individual*...........260
 *Throwing An Exception*...........265

## Metaphor – Levels Of Modelling...........266
 *Flow*...........266
 *Labels*...........267
 *Metaphor And Epistemology*...........267
 *Values And Beliefs*...........268
 *Problem Metaphors*...........268
 *Changing The Metaphor Uncovers The Deep Structure*...........269
 *Leaping Between Metaphors To Access Flow*...........269
 *Metaphors And Consciousness*...........270
 *Perceptual Positioning*...........271
 *Gregory Bateson's Learning Levels*...........271
 *The Matrix Metaphor*...........271
 *Problem Solving With Metaphors*...........272
 *Decoupling The Metaphor*...........274
 *Riding The Metaphor*...........275
 *Vertical And Lateral Thinking About Metaphors*...........276
 *Semantics*...........277
 *Everything Is A Metaphor*...........277

## Concentric Contexts...........279
 *Knowledge Is Infinite*...........279
 *Enclosing Levels*...........279
 *Reality Keeps Changing With Perspective*...........280
 *The Graves Levels*...........281
 *Ubiquitous Levels*...........281

## Methodological Conclusion...........283
 *TOR – Terms Of Reference*...........283
 *Terms Of Reference Are Initial Conditions*...........283
 *Methodologies And Myths*...........284
 *The Three Levels Of Focus*...........284
 *The Most Sacred Of Sacred Cows*...........285
 *Business Relies On Society*...........287
 *Denominalising The Sacred Cow*...........288
 *Denominalising "What"*...........288
 *Denominalising "Why"*...........288
 *Denominalising "How"*...........289
 *The Evolution Of A Crime Against Reality*...........291
 *The Frontier Of Knowledge*...........296
 *Levels Of Elevation*...........297

## Framing Up...........299

    *13 Billion Years, 10,000 Galaxies*.................................................................*299*
    *Accelerated Through Our Own Brain*.......................................................*300*
    *Communication Is Feedback*......................................................................*301*
    *Data Compression*......................................................................................*301*
    *Approximation*............................................................................................*303*
    *The Mirror In The Mirror*..........................................................................*303*
    *We Send Ourselves*.....................................................................................*304*
    *The Practical Nature Of Love*...................................................................*304*
    *Love And Project Management*.................................................................*305*
    *Heaven Or Hell*..........................................................................................*306*
    *The Inescapable Conclusion*......................................................................*306*
    *Polishing The Mirror To See A Better Reflection*....................................*308*
    *Balance, Predictability And Resolution*...................................................*309*

**Consequences**..................................................................................................**311**
    *The Horse*....................................................................................................*311*
    *Reality As The Lever*..................................................................................*312*
    *Humans As The Fulcrum*..........................................................................*312*
    *The Mathematics Of Change*....................................................................*313*
    *The Dog*.......................................................................................................*313*
    *Complex Numbers, Complex Systems – Reality Is Curved*.....................*315*
    *Boundaries*..................................................................................................*315*
    *The Slave*....................................................................................................*317*
    *The Truth About Happiness*......................................................................*318*
    *The Truth About Progress*.........................................................................*319*
    *The Truth About Processes*.......................................................................*321*
    *The Truth About Techniques*.....................................................................*322*
    *The Truth About Business*.........................................................................*322*

**Afterword**........................................................................................................**324**
    *Maintain Your Nerve*.................................................................................*324*
    *Maintain Your Perspective*........................................................................*324*
    *Maintain Your Vision*.................................................................................*325*
    *Maintain Appreciation Of Chaos*.............................................................*325*
    *Maintain Your Reality*...............................................................................*326*
    *Maintain Your Self*....................................................................................*326*
    *Maintain Your Life*....................................................................................*327*

**Appendix**..........................................................................................................**328**
    *People Quoted At Chapter Beginnings*....................................................*328*

**Index**..................................................................................................................**338**
**Bibliography**....................................................................................................**346**
**Author Online**.................................................................................................**349**

# List Of Illustrations

Ends and Means grid.............90
Tangential angles..................104
Relative infinities...................105
Paper plane...........................106
Rotating plane.......................106
Reality shells.........................108
Energy Horse........................109
Evolution of Knowledge I...121
Evolution of Knowledge II..122
Evolution of Knowledge III.123
Rational Magic......................127
Rational Movement...............128
Pre Motor Potential...............128
Delaying the conscious
    decision............................129
Curious Water.......................135
Three points of Triangulation
    .........................................147
Circle to sphere.....................151
2D representations rotate into
    3D....................................152
Science, Ethics, Morality......189
The Mirror of Versatility.......194
Slicing Reality.......................206
Assuming the slices of Reality
    are coplanar.......................207
Narrowing the context of the
    slices of Reality .................208
Deleting the connections
    between the slices of Reality
    .........................................208
Linear Spacetime from A to B
    .........................................208
Folded Spacetime .................209
Collapsing the paradoxes....209
Filtering the slices of Reality
    .........................................213
Intersecting the slices of reality
    .........................................213
Connecting the slices of Reality
    .........................................213
A simple feedback loop........216
Single Cause: Single Effect. .216
Associating cause and effect by
    proximity...........................218
Multiple Cause: Multiple
    Effect (Sequential).............219
Multiple Cause: Multiple
    Effect (Nested)...................219
First In First Out Queue.......220
Last In First Out Queue.......220
Multiple Cause: Multiple
    Effect (Random).................221
Single Cause: Multiple Effect
    .........................................223
Multiple Cause: Single Effect
    .........................................223
Single/Multiple Cause:
    Multiple/Single Effect........224
Cause and Effect Complexity
    Within A Frame..................224
Cause and Effect widening the
    Frame................................225
Mandelbrot Equation...........230
Mandelbrot Equation
    Expanded..........................230
The consequences of chaos....23
Concentric Realities.............280

# Section Zero: Overtures

# The Music Of Change

> *"Simplicity is the final achievement.*
> *After one has played a vast quantity*
> *of notes and more notes,*
> *it is simplicity that emerges*
> *as the crowning reward of art"*
> **Frederic Chopin**

Chopin, exquisite Chopin. He changed the perception of what could be played on a piano – forever. He wrote a handful of works, comparatively speaking. He was often ill and he was certainly a craftsman. His nocturnes are the pinnacle of an art. They are beautiful to listen to. They are technically stunning. They are ethereal yet grounded[1].

In the basement flat we sipped hot chocolate and listened to Chopin. Outside the rain fell on the London streets, as inside the notes fell on closed eyelids and open ears.

One piano, on a bare wooden floor, a cold concert hall ago, a nocturne captivated me for the first time. Playing on my stereo, a lifetime later, catching the mood of the firelight and the wine, it equally captivates me.

---
1  "His music is the universal language of human communication. When I play Chopin I know I speak directly to the hearts of people!" Arthur Rubinstein from the sleeve notes to *Artur Rubinstein -The Chopin Collection – The Nocturnes,* my personal favourite Chopin recording and the record referenced in the anecdote. http://www.amazon.com/Artur-Rubinstein-Chopin-Collection-Nocturnes/dp/B000003ENY

Each note is crystalline perfection. I picture this young man chipping away at the structures and bonds. He is testing, filtering and building to the ineffable will of what had to become, once it was begun.

At the keyboard the music is taking shape under his long fingers. He seeks the elusive combinations that will allow the energy to flow from end to end.

He is refining, reconnecting and realising, as the structure is built one passage at a time. The result is all that it needs to be, no more, no less.

It is as if the work is intertwined and intersected with existence. The listener hears the resonance inside. It echoes through the caverns of the brain.

From the infinite combinations that can stir the mind, Chopin chooses these notes.

The hot chocolate is scenting the air. The notes and the rain are falling in unison. It is a magical Saturday morning. We had turned from programming to music during a code review in a London basement flat.

Friday's office had been at an impasse. Saturday was ours and we had agreed to meet. The context shift could break the hold, we thought.

As the last notes resonated and the chocolate cooled, we somehow decided to put on some more

Chopin and brew some more chocolate. No more talk of work. We let our thoughts relax.

Monday morning was inspired. The problem was resolved before coffee break, in one of the infinite combinations of ways we knew.

# Contextualisation

We are lucky to be living through a singular revolution in the exploration of consciousness. For about four hundred years there has been an accepted model of how we interact with our brain. This has been the model of dualism. It is being re-thought and indeed re-taught.

The work and courage of some amazing scientists is challenging our perception of reality. The research in this field is as rich as it is ongoing.

You do not have to understand the deep mathematics, the complex psychology or the medical terms in order to start working with what these amazing people are telling us.

Certainly if you manage or interact with people, particularly people who use their brain for a living, it has become vital to understand the impact of this knowledge.

Every single area of human existence is being affected. Every single thing that every person does will benefit from this new perspective and the understanding that it is bringing.

There is a strong possibility that future generations will look back and see this age as a pivot point in the history of Humanity and social evolution.

My research in preparing this book has astounded, astonished and inspired me. I am convinced by the beauty of the human spirit. I am convinced that it is alive and well, despite all that history, ridicule, atrocity, fear and ignorance have thrown at it.

Everything we have ever achieved has been the result of someone daring to imagine it. They braved the wilderness to find the resources to turn a dream into reality. This is how great things are achieved.

# How To Read This Book

> *"My body has certainly wandered a good deal, but I have an uneasy suspicion that my mind has not wandered enough."*
> **Noel Coward**

The mind is made up of the conscious and the unconscious. We will explore both as the book progresses. There will also be talk about recursion and feedback loops. This book is an example of both.

The conscious mind deals in facts, logic and sound arguments. The unconscious mind deals in symbols, dreams and metaphors.

I have written both this book, and its predecessor, with these two constituent parts of the conscious in mind.

When both of these aspects of the mind are in a state of congruence, creativity is the result. This is expressed in many ways including: beauty and elegance, order and chaos, freedom and discipline, inspiration and appreciation, motion and rest.

We can call this 'flow', 'superconsciousness', 'peak experience', 'integration' or 'genius'. We all know when we achieve this state and we know what facilitates it. At those times we get a glimpse of what it might be like to be a Chopin or an Einstein. We glimpse the beauty of the possibilities, of the peak state, of where evolution can lead human beings if we stay the course.

Allow your conscious mind to be patient with the metaphors, images and recursion that I have interpolated with the material with which it more readily identifies. Ask it to reach out and embrace your silent, savant twin, with patience and kindness.

It has been shown by great teachers, like Michel Thomas, that great insight and rapid learning are associated with talking to both the conscious and unconscious simultaneously. It is enhanced by paying attention to feedback.

I understand that it is a strange recursion. I am asking you to allow me to use the subject and conclusions of the book to tell you what the book is about. This is because I am expecting you to know what you are going to read in order to read it.

It is like using Morse code to write a book about Morse code. It works because the unconscious mind already understands the code. Your unconscious already knows what I am talking about. That is the point.

If you do not immediately see the connections I am making, trust your own intuition. You will make the connections. We often learn something, or make a connection, that seems to race back through time and make sense of things we thought we did not understand. We can find ourselves reaching a deeper understanding and evolving our knowledge. The ability to change your mind is your prerogative and your strength.

You have an amazing mind. It surfs on a wave of time. It allows you to predict, plan and understand. As the penny drops, make sense of what used to be yesterdays mysteries. To be human is to constantly make sense of everything that happens. It exposes resources and wisdom that you did not know you knew you had.

The book is split into sections. Each of these has a short introduction which serves as a calibration or a focussing lens. It describes the shape of the idea for the section.

# A Nested Koan

## Let Information Flow Through Chaos

There is a pattern as we move through the chaos of existence to filter useful information out of raw data around us.

## Integrate Feedback With Connections

Are you open to making the connections that create the feedback? Do you listen to the connections that the feedback itself creates? It is through this feedback that we learn to make important and educated distinctions.

## Find Choice Where There Are Intersections

Infinite varieties of choice can be found at the intersection of boundaries.

## Examine And Refine The Loop

Solutions are the challenge that faces us at the beginning of every loop. We need to remove certainty and replace it with acceptance and learning.

---

*A koan is a question or a riddle which encourages you to take an unusual perspective. This book is intended to be a koan. This page is a calibration and a starting point.*

# Section One: Perpetually Becoming

# First Calibration

## It Is Through Doing That We Become

The evidence[2] that is currently emerging from Neuroscience and Genetics is fascinating. It suggests that the mechanisms of consciousness are more predictable and capable than had ever been suspected. It seems that our greatest weaknesses and greatest strengths as human beings spring from the same source: our ability to learn and adapt.

**Neuroscience** is the study of the nervous system. On the frontier of knowledge, the discovery of **neuroplasticity** is telling us that we really are what we do. We can change the very architecture of our brain and its abilities by our actions. Contrary to previously held convictions the brain is more adaptable than anyone had ever thought possible. Not only is it potentially adaptable, but it is actively adapting all the time as part of the process of living.

Critical periods which bring about deep changes in the structure of the brain, once thought to end in childhood, are now shown to be lifelong. You can, it appears, teach old dogs, new tricks.

---

[2] There is a lot of it around at the moment and it is explained in ways that the layman can understand. We will refer to accessible resources as we go.

**Genetics** is the study of the mechanics of hereditary transmission and variation in living organisms. It is the study of our DNA. On the frontier of knowledge, the discovery of **Epigenetics**[3] tells us that what we do changes our physical and mental abilities. We do not have to wait for evolution to change our DNA. Our actions and our environment change the genetic markers that decide how to express or activate the genes in our existing DNA. This happens in real time, meaning we are not as limited by our genetics as has been believed up until quite recently. The evidence suggests that very few experience their true potential.

Epigenetics studies the chemical markers derived from environment and behaviour. They can affect the activation or deactivation of genes in ourselves, our children, grandchildren and great-grandchildren. The sins and virtues of the fathers and mothers are truly visited upon the sons and daughters.

Nature or nurture is no longer the issue. It is all about environment. Because we can control our environment, by what we do, what we think and what we say, we can control our development. Knowledge is

---

3  Changes in heritable gene expression other than changes in the underlying DNA sequence.

telling us that we can indeed choose how we play the hand we have been dealt by nature.

IQ, genetics, wealth, upbringing, education and social status are not constraints. They are boundaries. Boundaries are beliefs and decisions. Although they are often confused, there is a world of difference between a boundary and a constraint[4].

If we are what we do, and we become what we pretend to be, then we have an urgent requirement to examine the way we work, live and play; what and how we think; and the things we subject ourselves to in the name of responsibility, entertainment and ambition.

While we wait for science to catch up with itself and deliver the great gifts it promises, we must use these revelations to evaluate the tools and choices we already have.

Some of it will confirm old ideas and some of it will confirm that some approaches, more than others, play to the strengths of our brains and our nature.

Ideas like Agile, NLP (Neuro Linguistic Programming), TOC (Theory Of Constraints), lateral

---

[4] E.g. A national border is a boundary, a mountain range or a sea is a constraint. One is notional and the other is existent whatever you label it. One changes with belief, perception and agreement while the other stays where it is whatever you decide about it. It is interesting to apply this logic to those things being presented as project or personal constraints.

thinking and emotional intelligence, by luck or judgement, seem to have tapped into the deep structure of reality. They may well be superseded as we discover more. While they seem to align with the surface structure of quantum physics, chaos theory, neuroscience and genetics, they are intuitive stabs in the dark that somehow hit the target.

As we wait for the lights to come on, we can:

1. Make use of what we know about the brain and ride the horse in the direction in which it is going.
2. Work with our nature rather than against it selecting the strengths bundled in our genes – which are considerable.
3. Evaluate the techniques we have available to us and use them wisely.

# Neuroprogramming

> *"We don't see with our eyes, we see with our brains."*
> Paul Bach Y Rita

## Discovering Neuroscience

Neuroscience is the study of the nervous system. It brings together the disciplines of biology, physics, engineering, philosophy, mathematics, chemistry, medicine and even computer science.

The Zeitgeist is full of books, films, articles and talks which will give you the background to neuroscience and brain anatomy. Even casual browsing will provide a cornucopia of detail.

There are ongoing discoveries about the adaptability of the brain. These are changing and refining the way we understand the phenomenon of consciousness. The advances in neuroscience are every bit as momentous as the mapping of the human genome.

All of this is available to us. We can no longer claim ignorance or insist that this is some sort of remote ivory tower science. This is the most intimate science imaginable. It increases us. It explains us. It offers us limitless horizons. It is immediate. It is exciting. It is challenging. It is awe-inspiring. It is happening now.

Regardless of the expertise that was required to explore, theorise, study, prove, peer review and publish this work, what it is telling us is really quite simple. What you do, and how you do it, shapes you and shapes your reality.

That we function like this is only one side of the story. What we can do with this knowledge is the other. We can begin to understand ourselves and one another.

## Neuroplasticity

We are all neuroprogrammers[5]. You cannot not change. You cannot not communicate. You cannot not rewire your brain.

The theory of brain plasticity is exciting. Discovering it is a revelation. It is one of those experiences that you will always associate with where you were when you realised what its implications are.

It is changing the way we think about ourselves and the way we think about what we do, how we learn, why it is difficult to effect change and how to overcome that difficulty.

Whether we knew about it or not, the brain has always been plastic. It has always been integrative and programmable. Knowing about it gives you choice and some measure of control.

Your brain is continually being rewired. Everything you experience is rewiring and conditioning you. This started while you were in the womb and will continue as long as you live. Luckily we have some choice, we are more than blank canvases for the unscrupulous to brainwash.

We are surrounded by many vested interests that will happily take over our neuroprogramming for us. These interlopers come under the guise of business, religion, advertising, fashion, television and a media that deals in convenient prejudice and "harmless" superstition.

Knowing about brain plasticity challenges us to stop in our tracks. It gives us the option to decide to stop being a hostage. It is more imperative than ever to choose to be your own neuroprogrammer.

---

5   I am commandeering this term to describe the control and responsibility we can assume over out own brain's plasticity – its ability to change.

There is a truism that all systems engineers know. RIRO[6]: 'Rubbish in, rubbish out'. If you supply even the best system with poor quality input, you will get poor quality results.

By understanding how attentive your brain is to input, and how easy it is to focus it on what you want, you can take the reins and dictate the shape of your own brain. Since this is the organ through which you experience reality, it is the most effective single point of change.

Neuroplasticity is the language of neuroprogramming, so it is important to understand some of its basic principles[7].

## The Important Principles Of Neuroplasticity

1. **Plasticity is a physical phenomenon.** The brain makes a physical change with every task we perform. Neurons wire and unwire. They associate and dissociate. They refine their function based on what we do and how we do it.

2. **Plasticity is lifelong.** The brain rewires itself constantly and finds new pathways to overcome even spectacular damage. It can adapt to massive change, whatever age you are.

3. **Plasticity is more powerful than genetics.** Thinking learning and acting can make changes at a genetic level by turning certain genes on and off[8].

4. **Competitive Plasticity.** The brain will use all available processing power. It does not waste. If you are not using a skill or ability, the brain will use this processing resource

---

6 This is also phrased as GIGO, 'garbage in, garbage out'. It was meant as a computer science pun on the term 'first in, first out', which, coincidentally, we also use later on, as an explanatory metaphor.
7 I have drawn these from multiple sources, scientific papers, articles and interviews which I will reference in the bibliography but the point of easiest access is "The Brain that Changes Itself" by Norman Doidge
8 See section on Epigenetics page 184

for something else. If you lose a finger or you go blind, the brain will remap and use the freed up, dedicated resources. This equally applies to playing the piano, programming Java or reading. If you want to be good at them you need to keep doing them. This is called competitive plasticity. It is either a huge liability or the most useful piece of knowledge you are ever likely to own[9].

5. **Focus:** What you concentrate on, and how you concentrate on it, determines what your brain considers to be its priorities. It will respond accordingly. What you concentrate on is what you get. You get what you concentrate on. Half hearted efforts get half hearted results. If you keep focusing on what you don't want, that is what you will get. This has a practical aspect when it comes to educating and learning. There really are short cuts after all. We know how to make our brain sit up and take notice.

6. **Efficiency:** The brain is constantly searching for the most efficient pathways. The more often you do something the more efficient the brain pathways become. This is a form of data compression that allows us to excel at very complex tasks by filtering out the unnecessary. We train our neurons to notice key information, whether this is tactile or mental.

7. **Association:** Neurons that fire together wire together and once they are wired they continue to fire together. Things that happen together, habitually, become physically associated into the same pathways in the brain. This has huge ramifications in terms of conditioned behaviour and skill acquisition. It has even wider implications for society in the treatment of mental disorders, work and education.

---

9  See sections on unconscious competence in volume one.

8. **Dissociation:** Neurons that fire apart wire apart. This is the corollary. Things that are done apart are physically dissociated into separate pathways in the brain.

9. **Sensory Substitution.** The brain uses the same sort of internal signals for all input. This means that it can be trained to accept input from one sense as input for another. Blind people can learn to see through touch. A cane becomes the extension of the arm. The feedback from a musical instrument becomes as sensitive as the fingertips.

## Doing And Thinking

We can easily see how our actions affect our state of mind. It is immediate. Think of how you feel when you are doing something you hate. Think of how you feel when you are doing something you love. Even the flow of time changes for us depending on what we are doing. E.g. The long wait in a hospital or the flashing by of the last few days of a great holiday.

What we think can also affect our physical processes. Being worried or stressed can make it harder to sleep or digest, for instance.

To every action there is an equal an opposite reaction. What we think and do affects our brain physiology. Our brain physiology affects what we think and do.

## Determinism Is A Choice

There is the huge matter of determinism and free will sitting somewhere at the centre of all this. As we come to terms with the machinery in our head, consciousness becomes more mysterious.

One way of thinking about it is to imagine that we have an internal switch that on one setting can make us a mechanical follower of programming an on the other a decisive programmer.

The robot[10] in our brain is our servant. It has our best interest at heart and is attuned attentively to our programming. It makes no decisions about the content of that programming, but follows it faithfully and trusts our judgement. It endeavours to please us and to take over maintenance and daily running of **"Me Ltd."** so that we can take care of executive functions. Its job is to free us up to search out new horizons and experience.

However, many things in modern society can set the switch to autopilot. The switch can get stuck in the robot setting. We can forget about the programmer state or even that the programmer state even exists.

What we need in our modern lives is a way to remind ourselves to throw that switch and wake up. When something does wake us up we need to recognise that as a resourceful state.

## Free Will Is A Habit

Questioning the habit needs to become a habit. We all have things that shock us awake and times when we become the programmer. In this state we look around and see the possibilities.

Unfortunately many people just grumble at the interruption and return straight away to their somnambulist stupor, letting the robot get on with it and operate on the basis of very flimsy data. We can mistakenly believe that the programming is set in concrete and that habit is the master rather than the servant. Even worse we can surrender control by conforming

---

10 Thanks to Colin Wilson author of "The Outsider" and "Superconsciousness" for this "robot" metaphor.

to those with vested interests who know how easy[11] the programming really is.

In our very language we refer to those, who are clever, responsible and resourceful, as being "switched on".

It is a useful tactic to take notice of those things that switch us on so that we can use them to retrieve those switched on states when we need them.

---

[11] Joseph Goebbels, Hitler's Propaganda minister, is infamously said to have claimed that if you tell people something often and simply enough they will believe it, no matter how outrageous it is. This combination of simplicity and repetition taps into what we are examining.

## Essential Things To Know About Your Brain

1. Your brain is capable of reorganising itself. It survives in a changing world by changing itself. This is a fundamental aspect of how it actually functions and deals with reality.

2. What you do shapes your brain. What you do is shaping you. What you do cannot not affect you. This law applies to everyone. No one has immunity. If it is not okay for others, it is not okay for you.

3. How, and what, you think shapes your brain. The thoughts you have regularly and how you think them affect the physical make-up of your brain. If you surround yourself with depressing and horrific input, that will shape your thought process and your brain. If you surround yourself with intelligent and resourceful input, that will shape your thought process and your brain. No one is immune.

4. The map of your brain is topographical but it changes. The physical location for things you use together, like your index finger and thumb, are mapped beside each other on your brain. How much real estate is given to processes, and where they are mapped, is constantly changing depending on who you are, what you are doing and how you are doing it.

5. Attention and focus make a difference to laying down new neuronal patterns. Multitasking is not such a great idea for certain activities, especially when you are trying to learn a new skill or take in information. You need to mark important things out to your brain by giving them your undivided attention.

6. What your brain perceives can be decoupled from the signal it gets. We are poly-sensory. Your brain perceives electrical signals and endows them with

meaning based on feedback and supporting evidence. Your brain can be persuaded to change its input channels. This is the basis of how cochlear implants and other such devices work.

7. Your brain responds to an environment that is rich and stimulating. If you want to acquire a new skill, it must be prioritised, presented in a variety of ways and made interesting.

8. If you want to be good at something you must keep doing it. If you do not use a skill your brain will use the real estate for something else.

9. Your brain uses immediate feedback to determine which things are important and which things should be connected. If you want to learn, you must put cause and effect close together.

10. Bad habits can take over brain maps. This is related to competitive plasticity. The brain gives space to what you flag as important. You flag something as important by doing it a lot. This means that we have to unlearn bad habits so that we can free up the resource they are using. (If you think about it, this is logical. You develop bad habits through giving them time and effort. Your brain has been wired to think that the things you give this time and effort to are the priorities. These things undoubtedly had some perceived secondary gain when you initiated them. Unfortunately we often hide that secondary gain[12] from ourselves and are left with the legacy which we must deal with. )

11. When you do new complex and difficult things your brain becomes more resilient to damage and more flexible. Learning is physical. It is a process of **synaptic connection,** the creation of new connections between brain cells, and **neurogenesis,** the creation of new brain cells. As well as creating new

---
12 Secondary gain: See the chapter "Useful Concepts" on page 182

connections between neurons, learning promotes the creation of new brain cells. This happens no matter what age you are[13]. This has come as somewhat of a surprise to the medical and scientific community. Apparently as we learn difficult things (e.g. languages and games like chess) or things that combine the mental and the physical (e.g. dance), our brain becomes more resilient. It is even thought that this may provide a way to battle Alzheimer's disease. Studies, particularly the "nun study"[14] carried out by David Snowdon in 1991 and continued to the present, have shown that people who have better linguistic skills, enriched, active, stimulating environments and education levels, can be asymptomatic even if they have the pathology of the disease. In addition they live longer, active and healthy lives.

It seems that highly developed neurons respond better to intellectual enrichment. We could say that the elderly actually have an advantage when it comes to learning new things.

12. The brain engages in specialisation and growth simultaneously. I like to think of it as a form of data compression. When we are learning to do something we throw large amounts of neurons at it. When we are learning to do something new, like play a musical instrument with a finger for example, the size of the brain map for the finger grows. As we become proficient the neurons required to do the specifics of the task become more selective. Only the appropriate neurons are used. The feedback mechanism becomes more sensitive and the selected neurons do more. This means that we have more sensitivity. Each neuron

---

[13] http://www.sfn.org/index.aspx?pagename=brainbriefings_adult_neurogenesis
[14] The Nuns study
http://www.time.com/time/magazine/article/0,9171,999867,00.html
http://www.youtube.com/watch?v=nw2lafKIEio

controls a smaller and smaller part of the finger. The map becomes more detailed.

13. The speed of thought is plastic. The more you think the faster you think. As we train neurons, they fire faster and faster and they need less rest between firings. Speed of thought is a critical factor in intelligence. You can change your intelligence level by thinking. It may seem like a truism, but it challenges the whole idea of genetic IQ levels. We are not limited by our genes. Far from it.

14. Memory can only be as good as the original signal. The ability to make finer and finer distinctions is therefore essential to learning and experiencing. We want to get more information from the original experience. The faster neurons fire, the clearer they become and the more they can synch with each other. When they work with each other they give a clearer signal. Training our brain to recognise distinction is therefore essential. In other words, we get a big payback from encouraging more neurons and training them to be faster neurons.

15. Culture shapes your brain. Culture, as in cultured activities, and culture, as in shared behaviour and norms of the group to which you belong. Music, art and literature make extraordinary demands on the brain. These things cause us to rewire in order to appreciate the distinctions. The ability to appreciate these distinctions expands our ability to appreciate experience itself. Engaging in these activities as musicians, artists, or engaging in any complex activity, causes the maps in the brain to enlarge for the resources we are using. This is sometimes referred to as perceptual learning, i.e. learning that changes our perception.

16. We are influenced by culture and other brains. Any sustained activity affects the plasticity of the brain.

Language, customs, religious beliefs, how we interact and what we observe the people around us doing, all affect the wiring of our brain. Our brain contains mirror neurons[15] that seem to allow us to experience the world from different perspectives and to allow us to learn quickly by example and through our environment. The explanations, that the research into these curious neurons offer, appeal to our common sense. Our environment and our long term sustained activities both affect our brain anatomy, which affects our perception and our abilities.

17. **Plasticity has a dual nature.** This is a warning. Plasticity has a paradox at its centre. It is a theme of this book that in order to be very useful, things tend to have to be very flexible and versatile. A knife can be the tool of a surgeon or a murderer. This can give very powerful things the potential to have equal and opposite effects to the ones you desire.

    The same resources that make the brain plastic can be used to make the brain rigid. The same features of the brain which allow plasticity by responding to stimulus, also respond to routine and lack of change. The brain can become stuck in a rut. If you do not feed it and use it and surround it with a rich environment, it adapts to that reality. Lack of mental flexibility is not a function of age; it is a function of habit. We notice that old people lose mental flexibility and we assume that age is the culprit. This is a complex equivalence. We see something and leap to the wrong conclusion based on circumstantial evidence. As has been mentioned:

    a. The brain continues to carry out neurogenesis until the day we draw our last breath. Even if you live to be a centenarian and beyond.

    b. More highly developed neurons, i.e. older neurons, respond better to learning.

---
15 Mirror Neurons: see "Mirror Neurons" on page 187

So, beware of mind deadening environments, boring activities, thought killing processes and bad habits.

## Summary

- The brain is constantly changing. This is as unconscious as breathing. Like breathing it follows certain principles. You cannot consciously control the functioning of the alveoli of your lungs but you can make sure they stay healthy and grow. If you do certain things like breathe deeply, exercise and avoid pollutants, they will respond predictably. The brain, likewise, responds to our actions, thoughts and habits. Neuroscience appears to be uncovering useful principles that we can use to challenge certain limiting beliefs.

## Conclusion

- Our brain is shaped by everything we do. Our consciousness resides in our brain; therefore our interaction with reality is affected by everything we do.

- Brain plasticity, and our ability to adapt to almost any environment, is evolution's greatest gift to us.

# Mapping The Little Grey Cells

*"He has awakened in me*
*a passionate consciousness*
*of the significance of life"*
**Siegfried Sassoon about Dr W.H.Rivers**

## Neuro Progress

Because of the importance and value to the rest of us; and because they want us to understand and use their findings, neuroscientists are keen to educate and elucidate.

### *Fear of the subconscious*

This has not always been the case. For a very long time it was feared, particularly by anyone who had studied Sigmund Freud, that the mind, particularly the unconscious, was a dangerous place, full of repressed desires and dark throwbacks to our evolutionary past. Consciousness was considered to be too dangerous to explore without the proper qualifications and safeguards. Things might get in and influence us. Things might escape and betray us. We might unleash nightmares.

Neuroscience is overturning all that. It is telling us that, quite simply, our unconscious, instead of being a lurking monster, is a slumbering giant of resource and creativity.

Many of the nightmarish fears are probably nothing more than just that – fears. It seems that we are not lumbered at birth with bestial tendencies that need to be repressed under a thin veneer of civilisation. It seems that we are born with a pressing and pragmatic desire to learn.

The deep dark monsters spotted in the unconscious, or as it can be more eerily labelled, subconscious, are shadows. They have no substance other than that which fear and ignorance give them and which learning and enlightenment can dispel.

If you tell your brain that it must have Oedipal or murderous yearnings, often enough, it will do its best to conjure them up for you. At first they will be shadowy formless shapes because your brain does not have enough detail. In time, with enough information about what is expected, it will do a fine job of fleshing them out.

Never has it been more true that you find what you look for. We are not naturally bestial, selfish, violent and repressed. These traits are learned behaviours. Neither are they in our genes waiting to pop out like some scientific fatalism.

The brain is an amazingly versatile reality translator. It makes the best of what you feed it in order to present you with a reality that matches your expectations.

### *The brain is a learning machine*

The brain is a voracious learning machine. If you do not give it the information it needs, it starts to make it up.

Censorship, fear, secrecy, dictatorship, repression of ideas and attempts to discourage enquiry, always have this problem. What people make up is always more dangerous than the truth.

Therefore the rational thing to do seems to be to give input and information to growing minds. Since we have seen that growth is the natural state of the mind, rich stimulating environments are best for everyone.

If you are running a state, a community, a family, a school or a business, free enquiry and access to high quality information would seem to be the most sensible way to proceed. It is the only way to avoid the creation of lurking monsters, paranoia and conspiracy theories.

It is tempting to argue that the fear and the monsters may control people or frighten them into behaving. This may even work for a while. The problem with imaginary monsters is that they are hard to contain and you cannot negotiate very well with them.

## *Localisation*

The Medical and scientific community have known for some time that the body and the senses are mapped onto the physiology of the brain. It was a long struggle[16] to discover this and to have it accepted.

The theory came to some understandable but incorrect conclusions that went unchallenged for centuries. It asserted that these brain maps were the same for everybody and that they were fixed and localised. The theory was that this mapping was unchangeable after certain critical periods in childhood.

It was also logical to infer that if certain parts of the brain were fixed to certain functions, that when those parts were damaged or underdeveloped, the function would be irretrievably impaired. This was accompanied by the theory that intelligence and ability was set down and limited at birth. The discovery of genetics seemed to support this. Because of these powerful and, seemingly, logically unassailable ideas, there has been a huge inertia against the idea of brain plasticity.

## *Challenging Localisation*

The story of how this view was challenged and changed is fascinating. It is full of quite astonishing people breaking out

---

16 The story of the discovery of the speech centres of the brain by Paul Broca is beyond the scope of this book but well worth reading about. His biography by the neuroscientist Francis Schiller is: "Paul Broca: Founder of French Anthropology, Explorer of the Brain". There is a good summary of his achievements and life at http://www.whonamedit.com/doctor.cfm/1982.html

of constrained thinking. Much of this has caused great discomfort and upheaval. It is providing hope and explanations as to why many effective treatments, considered to be unorthodox, work. It also highlights why some orthodox ideas are mistaken and are based on coincidence. Hundreds of years of accepted practice and theory is proving to be incomplete and some of it just wrong.

There are wonderful books that explore this and describe breakthroughs in the treatment of almost everything connected to the brain, from stroke rehabilitation to autism.

While this is fascinating we are concerned here with the effect on how we live and work. What we need to know is that the evidence for brain plasticity has become an overwhelming tidal wave.

## Finding out more about the history of Neuroscience

In the bibliography I have given a list of books and throughout the book there are internet resources which make this subject accessible to even those of us who are not psychologists or neuroscientists.

A small amount of investigation will uncover current and accepted scientific theory, backed up with empirical evidence and repeatable experimental procedure, which tells us that:

1. The mind body connection is a reality. How we think about ourselves and others really has an effect. We are self fulfilling prophesies.

2. We create and maintain a map of our reality as it suggests itself to our senses. This map exists on our brain and we interact with it. How we observe the world determines the map and the map determines the sort of world we live in.

3. We can recover from catastrophic damage and adapt to almost any environment – environments that require

us to conform shrink us and environments rich with intellectual challenge stretch us.

Before we start tinkering under the hood, let's have a look at the basic mechanics and the bodywork.

## *Brain Maps*

**The Visual Cortex** which specialises in visual processing is located in the occipital lobe at the rear of the brain above the cerebellum. It is the biggest single structure in the brain.

**The Auditory Cortex** which specialises in auditory processing is located in the temporal lobes on both the left and right side of the brain.

**The Motor Cortex** which specialises in planning and executing movement is located in the frontal lobes near the top of the brain.

However fascinating and tempting it is to explore the anatomy of the brain, this is not the place for an exhaustive tour. We just need to recognise that the brain has a general topography and that has been well mapped.

Each of these areas is split further into more specialised function. Everything we use to interact with reality is mapped to a physical location in our brain. This concept has a curious recursive feel to it.

## *The Penfield Cortical Homunculus*

Brain maps are most easily understood when we think about the Penfield Homunculus. You might remember seeing the image of the little man with great big hands and mouth and tiny body. It is a proportionally representative map of how much of our brain is used for what.

This map is like a pie chart of the brain showing how much of the motor cortex is given over to processing which sets of nerves. In addition it shows where these parts are on our brain in relation to each other:

Neuroscientists have mapped function to location in the brain with ever increasingly sophisticated and accurate tools.

In the pioneering days of Penfield[17] it was an instrument pressed on the brain during brain surgery. This excited the region and feedback was elicited from the patient. It measured clusters of thousands of neurons.

Today we have instrumentation sensitive enough to monitor the firing of a single neuron. We have tools such as functional Magnetic Resonance Imaging (fMRI), capable of analysing electrical activity in the brain in real time. This feedback loop has provided some very interesting results.

Experiments show that when people watch the real time activity of their brain as they experience pain, they can learn to control it[18]. There appears to be growing evidence of the connection between the ability of thought to affect the brain that is thinking it.

## *Scientific Inertia*

As with most things, the breakthroughs of one generation become the unquestionable reality of the next. Models again!

---

17 Wilder Penfield (1891-1976) Neurosurgeon known in his lifetime as the greatest living Canadian. He established the Montreal Neurological Institute in 1934. In searching for a treatment for epilepsy, he discovered that, by stimulating the temporal lobes during surgery, he could elicit clear integrated memories including sound movement and colour. He had discovered a physical basis for memory. He used the technique to map the physiology of the brain. He was a skilled surgeon who was praised by his peers as putting his patients before everything else. He wrote many books and the final one, published just weeks before his death at 85, "No Man Alone", was an autobiography stressing the importance of teamwork in neurological research.

18 Christopher deCharms has a study paper at
http://www.omneuron.com/PNAS_study.html

When the evidence began to come in that there was more to learn about localisation and critical periods, it was a challenge for people who had become entrenched in the existing model. They struggled to adjust their ways of thinking and working in line with the new knowledge. Progress is not always kind to those who have dug into a position.

Much science, psychiatric therapy and medical procedure rested on localisation being a fact. If, as has proved to be the case, it had been overstated, much therapy and medicine would have to be redesigned and much theory would have to be relearned.

The scientists and researchers who challenged localisation were ignored, ridiculed, unpublished and in some cases threatened professionally, legally and physically.

Science must seriously question any findings that change paradigms. This is particularly true when something has been accepted theory for generations and has had practical applications. However, the nature of science is to examine, to discover and to advance.

The scientific community eventually listened when it became clear that these new theories were based on solid research, repeatable results and sound reasoning. Even the most strident opposition is coming to terms with, and benefiting from, the new horizons that have been opened.

Localisation has been overtaken by knowledge and it has been accepted that while function is localised, it is not fixed. The brain is plastic. It is plastic for your whole life.

## *The Plastic Paradox and Critical Periods*

What had been observed as localisation was the plastic paradox. The abilities that make the brain plastic can also allow it to tenaciously hold on to learning and habit.

It is easier to write to a blank sheet than to one on which you have to compete for space. Many of these critical periods happen before the age of about eight. Experiences during the early critical periods were essentially writing on blank sheets. They did not have to deal with competitive plasticity in the same way later experience would.

The simplest metaphor I can think of is to compare the brain to computer disks. It was believed that it was a "write once read many"[19] disc with only one shot at formatting partitions and layout.

The brain is more like an expanding rewritable medium. The partitions can be moved and information can be augmented and moved around. We can see that the brain, even when horrifically injured, can find new routes to function, if enough healthy matter remains. This can be observed when stroke victims learn to use new areas of the brain to regain lost functionality.

Science is as sure of the plasticity of the brain as it can be of anything. Studying people with brain injury tells us just how adaptable it is. The ability to effect change within the brain does not disappear with childhood, it just needs to be managed differently.

It is possible to change the topography, but you also have to deal with the inertia of the habits and mapping that has already been laid down. The same can be said of science and the scientific community, which, unsurprisingly, mimics the brain and body in its behaviour.

---

19 Write Once, Read Many – data discs that could be written to but not written over again. If you use a CD or DVD in your PC they come in two basic types – once of which is WORM. There are discs that you can write data to but not delete it from. You can set it up so that you can write data to it in various sessions. (ok I know CD's are not strictly WORM, but it is a metaphor).

## The mind/body opportunity

Professor V.P. Ramachandran talks about the logical conundrums that confuse us when we think about how we see[20]. We assume that we have an image transmitted along the optic nerve which is viewed by some entity or, as he puts it, "little chap" in our brain who looks at the image.

Neuroscience tells us that we don't have a little chap in our brain. Even if we did, we would still have the problem of the little chap in his brain and so on, like turtles all the way back. There are no little chaps. Our brain decodes electrical signals and gives them meaning within a model stored therein.

The mind is an evolutionary function of the brain. That does not make consciousness any less amazing or mysterious. It makes it more astonishing and interesting. It simply allows us to harness its possibilities. It makes consciousness something we can study, question and dare to explore as part of the natural world.

## Opening a new frontier

Brain imaging, scientific evidence and an increasingly educated society are all helping us face and celebrate our status as biological beings.

There have always been people who dare to investigate accepted wisdom and demand proof. They ask questions and investigate complacency, superstition and tradition. They used to be burned as witches. Nowadays they face up to the scorn and scepticism of their peers. They are rarely recognised until they have flown so close to the flame that they cast the shadows of giants.

---

20 He gave the 2003 Reith Lectures and they are still available at http://www.bbc.co.uk/radio4/reith2003/lecture2.shtml or in book form "The Emerging Mind: The BBC Reith Lectures 2003" by Vilayanur Ramachandran.

Reading the accounts of scientists who followed where brain research has led them is like reading some kind of epic Promethean tale.

It is a tale of human beings striving to understand the universe with the best tools they have at their disposal.

Here are just a few of those people:

**Paul Bach Y Rita** – who discovered Sensory Substitution. He proved that the brain is adaptable enough to start processing tactile messages from the skin as visual images and messages from an electrode on the tongue as balance information that can replace information from a damaged vestibular function in the inner ear.

He showed that there are no specialised messages in the brain, just messages that the brain has chosen to interpret as sound, vision and touch and that these can be interchanged. After investigating what he achieved it is impossible to see synaesthesia as anything but an ability. His work broke the mould and demonstrated brain plasticity beyond a shadow of a doubt.

**Michael Merzenich** – whose work combining medicine, neuroscience and engineering led directly to the development of a workable cochlear implant. Building on the work of Bach Y Rita he realised that it was not a question of inventing a sophisticated hearing aid, but of developing an instrument that takes advantage of the brain's amazing plasticity.

Among his many other achievements he may also have discovered both the cause and treatment for autism. He discovered that dyslexia comes about when there is an environmental or physical problem that stops a developing child from hearing and processing distinctions in linguistic sounds.

In treating dyslexic children he noticed that the treatment was also alleviating autistic symptoms. It seems that the inability of the brain to process distinctions in any sense affects the brains processing ability across the senses because the distinctions are in brain messages not sound, sight or touch. He has shown promising results when dealing with autistic children in his FastForward initiative which concentrates on restoring this distinction process through the brain's plasticity.

**Vilayanur. P. Ramachandran** – who leads the field like some scholarly detective. He calmly solved the mystery of phantom limb pain with a cardboard box, a mirror and a moment of shockingly obvious logic.

If you want to start understanding what consciousness is all about, it is worth seeking out and devouring every word this astonishing man makes available[21].

## *The adaptable brain*

The brain is far more adaptable then anyone suspected.

What we do frequently and habitually changes the physical wiring of the brain.

Think about this for a moment. What you do frequently and habitually as your everyday activities changes the physical organisation of your brain.

The implications are stupendous. It is reciprocal because the physical mapping of your brain dictates what you habitually do.

We are all affected by our environment. Environment and habit shape who we are. This happens, not on some mystical other mental plane, but in measurable physical changes to the brain in the cranium. We can change what our physical brains are capable of.

Contrarywise, as the Tweedle dum would observe, this also explains why it is so hard for some people to effect change.

---
21 "The Tell Tale Brain" by V.P.Ramachandran

Let me tell you about taking my children out to enjoy the snow.

### The Hill

The field next door had a large hill in the middle of it. The first time it snowed, we took the two children over to try out their new sleds.

We climbed over the fence. My wife waited at the foot of the hill to play the catcher in the rye, while I trudged up to the top with the girls. The expanse of white snow was quite beautiful.

The first child was installed in her sled and let go. Nothing happened. The snow was powdery and was not all that slippery.

I had to run down the hill behind her pushing. I could feel gravity doing a lot of the work as long as I provided a little momentum.

Although the snow had blanketed everything to a depth of about 6 inches, I could feel some of the natural topography of the hill suggesting part of the route. It was possible for me to override these and guide the sled to the bottom by what I felt was the safest route.

The child in the sled was delighted and squealed all the way down. She wanted to go again, straight away. We walked back to the top following our original footsteps. It was instinctive to do this rather than infringe on the virgin whiteness with, who knew what, lurking underneath.

At the top the other child was happily making snow angels. If you have never done this, let me recommend it as a worthy addition to the things you can do to remind yourself you are alive.

Voyage-the-second was easier. We only made minor adjustments to the original trajectory. I still had to work a bit to keep us on course.

After descent three, my daughter was able to slide to the bottom without any help. After I had taken a go myself, and flattened the snow to something approaching ice, she was able to gain a respectable velocity.

She decided she wanted a longer slide, but I now found it impossible to chart another route to the bottom from the same starting point. The luge we had created now dictated matters. Also, since the ride was faster and faster, she was going a little further each time and approaching a nasty patch of brambles.

So, being resourceful anthropoids, we chose another starting point. This time took a more exciting and adventurous route to the bottom, using the same sequence of events.

We had to be careful when we came close to the first track or we were likely to find ourselves following the original route and thereby running the risk of ending up in the brambles.

As we crossed over it we had to be careful that we were not diverted from our new, improved ride into the old, shorter less desirable one. Each time we had to be careful to focus our attention on the new route until it became progressively less risky.

By that time it was getting dark and we went inside for hot drinks and to toast ourselves by a log fire.

The next morning we discovered that there had been a heavy fall of snow again. We resumed our frolics on the wonderful hill. We found that there were only traces of our activity from the day before, but that it was easier to use the tracks we had already used.

Even though we had decided not to use the first track to the bottom, we found that we could now extend our second one to include that part of the hill if we were careful. The new snow in the groove made it less likely that we would get pulled into the rut.

After a few more days of snowfall and care to stick to the better luge, the original rut disappeared and we were able to really speed to the bottom with ease and safety.

The brain is much like this. You can choose any course of action but the more you do something, the harder it is to find or consider an alternative.

From all this I have derived a set of ten steps for change:

1. Get some idea of the underlying topography and potential risks before you start.

2. Use available resources and go with the flow until you know better.

3. Pay attention to the route and the destination – both are important.

4. If you want to make radical changes it is easiest to start from somewhere that offers a fresh perspective. It is almost impossible to change route or destination at full tilt.

5. If at all possible, act when there is snow forecast[22] or make sure you know where you can hire a snow cannon[23]. You will almost certainly encounter traces of how you used to do it, no matter where you start from and no matter how new and radical the change is. You will need something to fill in the old tracks as you cut new ones.

6. Use time as your ally.

7. Focus on where you want to be not where you don't want to be. Focus on what you want, not what you don't want.

8. Not everything is knowable until you are doing it, so be sensitive to the undulations and forces at work as you

---

22 i.e. when conditions are optimal
23 i.e. knowledge generator

progress and find out which ones are helping and which are hindering.

9. The most direct route sometimes, although not always, hides unacceptable consequences.

10. Practice what you want to do until it becomes the norm.

And the extra free bonus rule, eleven for the price of ten:

11. Make sure you have something worthwhile do to while you give the new snow a chance to fall and present new opportunities.

Here are some concepts that we need to pack and keep handy as we sally forth into this snowy landscape.

## The Brain Is A Feedback Mechanism

This is the riddle, the question, the answer, the problem, the solution and the recursive heart of the matter.

The most startling evidence of this is Professor Ramachandran's amazing solution to the century-old problem of phantom limb pains.

### *The phantom limb*

Phantom limb pains had been impossible to treat because they affect limbs which have been amputated and there is no physical limb to treat. The real problem, he theorised, is twofold:

The last message the brain gets from the limb is one of immobilisation before surgery or of the trauma of an accident. After that there is no more feedback from the limb.

This causes a problem because, as we have said, the brain is a feedback mechanism.

Let's imagine an amputated arm. The brain sends out messages to the limb to clench against the paralysis or the pain. Normally one or more of the senses would report that the action had been successful and the brain would stop sending clench messages. However with no feedback there is no shut off so the brain sends more clenching commands and then starts associating pain with the action.

There is no arm so you would believe that there should be no pain. Up to this the problem of phantom pain was believed to have been one with the nerve endings in the amputee's stump and the obvious treatment was to amputate another little bit.

Professor Ramachandran, in the tradition of all great sleuths, asked some important questions. He wondered if this pain could be unlearned by providing feedback.

He put a mirror in a box with two arm holes so that the patient could look in the top and see two arms if asked to place his physical arm in one hole and his phantom arm in the other. The patient, thanks to the mirror, saw two arms. Professor Ramachandran then asked him to clench and release both arms. The patient's visual sense saw two arms clench and two hands relax. After ten years of agony the phantom arm relaxed. Professor Ramachandran had been right and the brain got the feedback it needed to let go.

Here is how you can test it for yourself even if you have two arms. If you get a false arm and place it on a table and place your real arm under the table, you can carry out a remarkable piece of neuroscience experimentation.

Get a friend to simultaneously stroke your arm and the false arm with the same movements. In a short while, even though you know your arm is under the table, you will start to identify the false arm, the one you can see, as the one that is sending you the sensory feedback rather than the arm under the table.

You don't even need to obtain a realistic false arm. Any object, from a cricket bat to a long cushion, works just as well. Even stroking the table itself can also produce some sense that the tabletop is sending you sensory feedback – because that is the action you see that corresponds to the sensations you feel.

This is a matter of being able to confuse cause and effect.

### *The brain determines reality*

Professor Ramachandran has made some other interesting discoveries. Moving his attention from amputees to other people with chronic pain he carried out a similar experiment. Once again he fooled the brain into thinking that a chronically painful limb was being moved about without pain. When the limb was actually moved the pain was gone. In other words there is at least some pain that is a learned response that can be unlearned.

Pain is generated in the brain as a result of sensory feedback and mapped back onto the body – this is contrary to the one way message system that was assumed since Descartes theorised nervous signals as fluids flowing from the pineal gland to the body. If you reflect on it, this is something that should change our relationship with pain and pain treatment dramatically.

I had an older cousin by marriage who preached "mind over matter" to me while I was growing up. I saw him hit his right hand with a hammer, drop the hammer and shake and blow on his left hand. When I asked him about it he told me he was fooling his brain into thinking that there was no pain. He said the trick was to pretend for all you are worth that you have hurt the other hand and then realise that it is not hurting. There will be no pain.

He advised me to treat all pain in the same way and to follow his example. He also recommended marmalade as a condiment for cheese which horrified many of the conservative eaters in the family.

He professed to be George Harrison's first cousin. I thought this was fantasy until investigation revealed that the Beatle had indeed visited his family one summer.

My adopted cousin was a highly loved and unique individual until his last breath. He was the type of person who would have just wiggled his ears and said "of course" to both the ideas of neuroscience and nouvelle cuisine.

Writing this has made me chuckle with an amazingly vivid memory of his personality and presence. It is as if I saw and talked to him only moments ago. Isn't it great when that happens? You suddenly uncover the treasure you have stored in your own memory and bring amazing people back to life in your mind?

Whatever your brain thinks is happening, whether your conscious mind agrees or not, is what determines reality for you.

We will follow this train of thought later. For now, just remember that the brain deals in feedback though not always the feedback you might expect.

### *The effect of stress on the brain*

Stress is an interesting thing. We have the ability to be stressed for a reason. We need to send vital resources to our fight or flight mechanisms. This should be a short term diversion of vital processing power. We have contrived to find very inventive ways to make it last and to convince ourselves that it is necessary to do so.

Neuroscience is confirming[24] what we all knew all along, stress makes people stupid.

Here is some information that may help you to decide to give up stress for good:

Stress interferes with the way the brain fires neurons. They still fire but they fire at a different frequency. Stress slows the brain cell's ability to form the connections that form memories and that use that information for decision making.

By examining the hippocampus we know that stress interferes with people's ability to learn, remember and to make decisions.

In other words, stressed people really are less likely to make good decisions. One could almost say you are guaranteeing they will make bad ones.

I understand what some organisations and philosophies hope to gain by putting people under stress. There is a train of thought that should be lying rusting on the side track of history and pointed out as a warning. People are thought to need the motivation of stress to perform. The opposite is true.

In the workplace, 'command and control' management styles cause stress. A bullying boss causes stress. Micromanagement causes stress. Tiredness causes stress. Mismatch between what is being asked and what is possible causes stress. (This is sometimes, euphemistically, called giving people stretch goals). Too little authority and too much responsibility cause stress. Fear causes stress. Working with people you don't trust or like causes stress.

Competition creates stress, especially if people are in fear for their job or their self respect. The carriages on the stress train of thought are full of fanciful notions. They insist that we must discover the best people and get rid of the losers. The

---

24 http://uwnews.washington.edu/ni/article.asp?articleID=45318

presuppositions are as staggeringly arrogant as they are ignorant.

These notions are having a party at huge expense.

In reality what we are doing is putting everybody under stress and restricting the ability of people to grow and restricting the ability of the best to perform.

Most of the knowledge work that people perform requires them to

    a) Learn

    b) Remember

    c) Make decisions

In business terms, stress makes people unproductive. This is exactly the opposite of the declared purpose of putting and keeping people under stress at work.

It is not just knowledge workers that are affected by stress. How vitally important is it for all of us to learn, remember and make decisions?

# The Storm Of Ideas

> *"The things being mentioned
> are not necessarily
> the things being talked about."*
> **Gregory Bateson**

## The Things Being Talked About

Whatever else we appear to be mentioning, we are always discussing feedback. This is the rock at the eye of the storm, the hero's talisman, the true name, the word of power, the ruby slippers that take you home, the Maltese Falcon, the maguffin in the suitcase, the singularity at the heart of the difference engine and the power that drives a single human heart to make a difference.

## The Brain Storm

I love writing. Words and computer code are ideas made visible. I sit and let my imagination run where it will and watch in rapt fascination as logic unfolds. It is like diving down a cliff whose face is alive with ideas, like sheet lightening, and thoughts like explosions of barely contained energy.

The air is full of potential, charge and electricity ready to be carried back to the surface. I often look down, amazed at the pages of words I have written, having only the vaguest memory of having written them.

Perhaps these are trips into my unconscious mind and perhaps the cliff is my own brain. I am fascinated by the idea of ideas and where exactly they come from. Are they a result of spontaneity or control? Are they rare, requiring the utmost care and attention to birth them into reality or are they limitless and requiring only that we relax and let them flow?

New understanding of the workings of the brain and of the pathology of consciousness seems to suggest both.

This storm of ideas is where we must go if we want to access the heart of our own creative resources. We must launch ourselves off that cliff and as we plummet we must turn our fall into a dive and then into stillness, flying, as we do in dreams, by rejecting the ground so that the cliff face rushes by and we inhabit a position in the singularity or our own consciousness.

# Shaping Reality

> *"Life is a whim of several billion cells to be you for a while"*
> Groucho Marx

## The Resilience Of The Brain

Brain Plasticity[25] is the ability of the brain to change and adapt. Plasticity means that it can undergo persistent change in response to experience.

One of the most moving, inspiring and genuinely jaw-dropping things I have ever seen was a short documentary film about a young girl who had the entire right hemisphere of her brain removed surgically[26].

I was amazed by this particular little girl but even more so by the implications about the equipment in my own head.

Before the operation she had Rasmussen's syndrome[27] which is a type of epilepsy. There was an electrical storm originating in the right hemisphere of her brain. She could not use her left arm and could just about use her left leg[28]. The fits were life threatening and totally debilitating. Surgeons decided the only way to proceed was to remove the right half of her brain.

The right hemisphere controls the left half of the body and the left hemisphere controls the right half of the body. You might expect that with the right half of the brain missing the left side of the body would be paralysed.

---

25 Also known as neuro plasticity
26 http://videos.howstuffworks.com/tlc/28671-understanding-brain-epilepsy-surgery-video.htm
27 http://www.epilepsy.org.uk/info/rasmussen.html
28 http://www.youtube.com/watch?v=TSu9HGnlMV0

Before the discovery of the extent to which our brains are plastic this concern might have discouraged anyone from carrying out this operation.

The surgeons knew about plasticity and went ahead, trusting in the amazing resilience of the brain.

Within ten days of the operation this little girl was walking and using her left hand. The brain is so plastic that, once the left side recognised that the right side was gone, it started rewiring itself – rapidly!

In the video you see her dancing and skipping. She is obviously a happy child.

The operation was ten years ago. In a follow up article in The John's Hopkins magazine, Dome, in 2009[29] she is described as "an effervescent, 13-year-old middle school student who earns straight A's and spends her spare time devouring Harry Potter."

This is just one of the amazing stories out there. There is the woman whose brain is hollow[30] but who has above average intelligence. There are people born with only half a brain who manage quite well thank you, and even rewire the input from both eyes to be processed in only one hemisphere.[31]

That the brain adapts and changes may seem like a self evident truth, but many other self evident fallacies rest on the assumption that the brain is not plastic. We subscribe to many of them daily without realising it. How many do you recognise and how many could you add to this list?

List of harmful lies about you:

1. It gets harder to learn as you grow older

---

29 http://www.hopkinsmedicine.org/dome/0903/centerpiece.cfm
30 http://www.mymultiplesclerosis.co.uk/misc/mysterious-brain.html
31 http://www.dailymail.co.uk/health/article-1200958/Girl-born-half-brain-person-world-fields-vision-eye.html

2. Your memory fails as you grow older

3. You are limited by the intelligence you were born with

4. You only use x% of your brain

5. You live in a rigid box of class, genetics and environment.

Take my advice and refute the lot. You have everything to gain. If the human brain is adaptable enough to cope with losing a whole hemisphere and to set about learning how to wire half the body to the remaining one within days, think about the potential for the ordinary tasks we ask it to learn?

## Summary

- Before the removal of the right hemisphere of a girl's brain, the left side of her body, which it controlled, was dreadfully affected by the electrical storm there.

- After the operation the left hemisphere realised the right hemisphere was gone and wired itself automatically to control both sides of the body.

## Conclusion

The brain is not hard wired. It is plastic, changeable and much more adaptable and resourceful than anyone ever suspected.

# Section Two: The Nature of Knowledge

# Second Calibration

> *"The spice must flow!"*
> **Frank Herbert**

## The Spice Of Life

Who am I? Where did I come from? Why do I do the things I do? How do I know the things I know? How do I remember? How do I learn? How do I grow? What are my limitations? What is my potential?

As far as we know the human brain is the most complex thing in the universe. It is certainly the only thing we know of that is self aware; that can ask recursive questions about how it can ask questions and that is aware that it is aware.

Imagine this mind. It is a tool that can be used effectively to do whatever you want in whatever context you find yourself.

Whether you are a gamekeeper, a bank manager or a musician, all of your skill depends on and resides in your brain in the form of information. Who you are depends on how you interact with that information.

The brain works on feedback. It thrives on it. It loves to compare and contrast feedback with stored information. This is both an opportunity and a challenge. You can censor the feedback as it arrives – and stagnate; or you can learn from it – and grow.

People who censor the feedback limit their opportunities for growth. People who connect with and embrace the feedback build larger spaces for their own existence.

All this touches on ego and a sense of self. Both of these are models. They are narratives. They are myths. They are pitons in the walls of existence. We need them but we must be careful we do not get trapped in them or behind them.

People who are put in sensory deprivation chambers or people with conditions such as transient epileptic amnesia, find out very quickly that the self is neither James's pearl[32] nor Hume's bundle[33], but perhaps both and neither at the same time.

Being ourself is an ongoing process of dealing with information. The self is potential, the self is capacity, the self is practice. Without memory and feedback the bundles and pearls evaporate into an altered state. The information must flow for the brain to grow.

Managing knowledge is about increasing your brainpower.

---

32 William James, philosopher (1842-1910), describes the "central nucleus of the self". Galen Strawson, philosopher (1952 - ), describes the self as a string of pearls or SEMSETs (Subjects of Experience that are Single Mental Things).

33 David Hume, philosopher (1711-1776), proposed that the self is a bundle of sensations associated with self. Bundle theory insists that an object is nothing more than its properties. The self is nothing more than a bundle of interconnected sensations.

# Managing Philosophy

*"In all affairs it's a healthy thing now and then to hang a question mark on the things you have long taken for granted."*
**Bertrand Russell**

*"Art is the imposing of a pattern on experience, and our aesthetic enjoyment is recognition of the pattern."*
**Alfred North Whitehead**

## The Dragons Tail

Is there any place for philosophy in project management?

Philosophy comes from the Greek word meaning love of wisdom. Wisdom is defined as making the best use of knowledge. Knowledge is defined as expertise, and skills acquired by a person through experience or education.

I think we have caught sight of the dragon's tail.

## Stalking The Dragon

When John Lock questioned the divine right of kings and Newton ushered in a new age of science, Voltaire, considered by some to have been one of the most intelligent men to have lived and the philosopher of the enlightenment, echoed the thoughts of Diogenes when he said "prejudices are what fools use as reason" and that "It is hard to free fools from the chains they revere."

When we limit our thinking and scoff at efforts to raise our vision beyond "the other game"[34], which denies all motivation but profit, we support a prejudice against our own true nature. When we embrace those things that pull us out of ourselves and widen our vision we can often find that our nature surprises and rewards us.

---

34 The last volume of the Trousers of reality will deal with "the other game"

It is prejudice that tells us that we must not care if others suffer in order for us to profit and that in order to succeed we must concentrate on beating the competition at any cost.

It is a ferocious lie. It is a vast lie. It is a vile lie. It spawns lesser lies and hides itself in the dark weaving fear and unhappiness, worry and distrust. It is the lie of the zero sum game. It is the lie that says that you must create losers in order to have winners.

Over the last decades there has been a changing attitude to work. We allow different pressures to govern us and we experience a lack of time for ourselves. The workplace is demanding longer working hours, paying less and having a detrimental effect on our health through stress. Manipulative techniques are used to get compliance. Teamwork and loyalty are being exploited and have become perverted. Society is creating company robots by using words like passion and commitment to harness our need for purpose. We have been caught in a trap of consumerism and we are being fooled by a mirage[35].

## A Flight Toward Reason

Is it inevitable that we are willing slaves, burying our curious and questing nature in compliance? Is it really the price that employers demand that we are reduced to gaming a system that rewards the guileful?

Voltaire would smile. He also said "Each player must accept the cards life deals him or her: but once they are in hand, he or she alone must decide how to play the cards".

---

[35] Madeliine Bunting's book "Willing Slaves" reflects upon this and echoes the thinking of, Bertrand Russel, G.K.Chesterton, William Blake and others. http://www.amazon.co.uk/Willing-Slaves-Overwork-Culture-Ruling/dp/0007163711

According to Thomas Paine[36], and the philosophers of the enlightenment, it is our right and our responsibility to apply reason and to question authority.

"It is necessary to the happiness of man that he be mentally faithful to himself. Infidelity does not consist in believing, or in disbelieving, it consists in professing to believe what he does not believe." – Thomas Paine

So how do we know what to believe or not to believe; which things to challenge and which things to support?

Generations of philosophers have advocated reason and logic. They also tend to advocate compassion. Compassion is a form of empathy. To have empathy with others requires you to have empathy with yourself.

---

36 Paine wrote "The Rights of Man" 1791-2 (championing representative democracy) and "The Age of Reason" 1794-5(attacking institutionalised religion). He influenced Abraham Lincoln, Thomas Edison, Bertrand Russell among others. He had a hand in the writing of the American declaration of Independence.

# Empathise With Yourself

*"No man is hurt but by himself."*
**Diogenes of Sinope**

## Control Groups

To start discovering knowledge we must know about the control group[37]. We must be able to gauge the background noise, what we consider to be the norm, so that we can make clear distinctions and draw valid conclusions that lead to knowledge. What is information and what is belief? What is ourself and what is wishful thinking? Where are we absorbing and where are we reflecting? Where are we distorting and where are we transparent?

To give or take feedback and to have any relationship with knowledge, we first have to begin to understand our own limitations and strengths. We must know ourselves so that we can learn to be our own control group. We – our values, beliefs and consciousness – are the background to everything we do.

This is probably the single most frightening thing anybody ever says. Know thyself! We search and search, we peel away layer after layer but nowhere is there any self to be found. This is because of resonance and calibration. We are searching for something that we think should stand out from

---

[37] A control group is a group that is not subjected to the theory you are trying to test so that you can calibrate the effect of change. The simplest forms of controls in scientific experiments are:

Negative – you do not expect the phenomenon to occur normally. The control group is isolated from the "difference" and one group is subjected to it. If both groups respond similarly you have a confounding variable – an x factor - and your theory needs more work.

Positive – when you expect a phenomenon to occur normally without the "difference" but you can still use it to test a phenomenon. You introduce different "differences" and measure the effects. This is analogous to benchmarking in the world of computing.

the background noise. The thing doing the looking is the thing being looked for.

Our model of the physical world is based on seeing, hearing, feeling, tasting and smelling. These all rely on being able to perceive difference.

As we have learned about the workings of the brain we can catch sight of a deep principle: sight, sound and movement are absolutely infeasible.

We need to filter out imperfections – blobs, lines, spots and even gaps – from our sight, so that we see a clear picture. We need to filter out the sound of the blood in our ears, the beat of our heart and various buzzings, tickings and distortions, in order to hear clearly. In order to move we must ignore both the finite and the infinite to swim through an ocean of time and possibility as if it were a perfectly ordinary thing to do.

## Human Difference Engines

Consider a wave form. Exactly in phase, two waves will amplify to twice their size. Exactly out of phase, and they both disappear. Consider a deeper examination of a wave. It is a disturbance against some sort of background. It can only be perceived because it is different to the background.

Seeing, hearing, feeling, tasting and smelling all rely on distinctions. Our existence is made up of measuring differences. We are less like computers and more like difference engines[38].

---

38 Charles Babbage is credited with inventing the first computer and he called it "the difference engine". It was built in 2002 - 153 years after it was invented! It seems that he was beaten to it in concept by J.H.Muller who wrote a book about it in 1796, There is an enthralling site here http://www.computerhistory.org/babbage/. The first computer programmer was probably Ada Lovelace. Her father was Lord Byron . (Yes, that Lord Byron). She got the job because of her training in mathematics and yet she is the genius that saw the possibility of taking the difference engine from numbers into the symbols of music, writing and beyond.

When looking for myself I am invisible because I am the background.

Many of us are alarmed at the thought of our inner selves, the real us, stepping out from behind the very careful camouflages we have set up. What we consider to be character traits may turn out to be nothing more than memes[39]. There is a very real possibility that there is no such thing as the self.

## Facing Our Fears

Babies are born with only two fears: loud noises and falling. These are survival traits and are really reflexes rather than fears. All other fears are learned. All of the things we perceive as vices or character flaws are a response to fear of some sort. That is a big statement. We must again consider cause and effect.

Is it possible that greed is a fear of poverty which mutates? Is selfishness a fear of scarcity which becomes a compulsion? What if megalomania is a fear of being weak?

So how do these fears get started and what does traditional psychiatry do about them?

Often people seem to be a little nervous that if they connect with their unconscious, it will reveal monstrous[40] things about them. The last thing they want to be is that self. They prefer the nice person who keeps the self under control.

Logically, even if the self were as different from the public avatar as we seem to think, it would only reveal very human things about us. Generally, we are not nearly as good at acting

---

39 A meme is a viral idea or behaviour. Richard Dawkins proposed the idea while searching for a way to talk about evolution and genes in "The Selfish Gene".
http://www.rubinghscience.org/memetics/dawkinsmemes.html
40 Of course there are psychopaths out there and people with genuine mental illnesses. There are not as many as you might think, so, for now, let us limit ourselves to ourselves.

as we seem to think. Many of the things we think we are hiding are things that other people like about us. These can be things that make us more approachable.

## Embedded Perceptions

I was talking to a friend about the fascination with alien abduction.

He had noticed that a lot of the images presented of aliens portrayed them as tall thin creatures with big eyes and indistinct features. It seems that many reports of alien experiments involve a fascination with the nether regions of the unfortunate captive. There also seems to be a theme of paralysing rays and bright lights.

"Could it not be some sort of half remembered memory of being a baby?" he asked me.

It does seem plausible. From a baby's perspective we are approached by large looming figures in the night that turn on blinding lights. Babies are fixated with eyes as part of their development. Adults probably appear as indistinct alien figures with enormous eyes who that them helpless and do strange things to them such as changing nappies and applying lotion to rashes. This often happens in the middle of the night when most adults are bound to be a bit sleep deprived and mechanical.

Is it not possible that we might be storing this strange experience and that we bring it out when we are bored, to frighten ourselves?

Events, whose context we do not understand, can become associated with misunderstandings and false perceptions.

When you were a child you might have come across busy parents and been told that you should go play outside. Life long "issues" of not being loved can be started through not understanding the context of an adult world.

## The Russian Doll Self

Does the seven year old you used to be ever creep up on you and make you laugh until you cry at something silly? We are like an us-shaped onion of all the people we have been.

> I recently watched two carers bring a six foot young man to the playground where my children were playing. He charged about having a great time on the slides, swings and climbing frames. Most of the children prudently got out of his way but all of us, especially the parents of the children, were quite captivated watching him have the maximum of fun to be had from a playground.
>
> Naturally, as we kept an eye on our children, we realised that the carers had an impressive relationship with their charge. It was obvious, from the way that they all related, that there was a very functional and affectionate relationship in place.
>
> The thought crossed my mind as to whether he was aware that he was different, that there was something wrong with him. He was so much bigger than the children.
>
> I was writing this book so I was full of thoughts about models of the world and limiting beliefs. I looked up at him standing on top of the climbing frame, crowing out his joy and delight on a fabulous spring afternoon.
>
> My model of the world and the model of those around me was that a normal adult should sit and be grown up, while the children revel in the weather and swing upside down in the pursuit of fun.
>
> I realised that if I was honest I would have to admit that I was terribly jealous of my friend up there. If he thought about it at all, he probably wondered if we knew that we were different.
>
> I was struck that it was I who had something wrong. I was concerned about how other people could think of me. I was pretending to be grown up when I really wanted to jump on the swing and hoot like Peter Pan. The seven year old was still in me and urging me to give it a go.

## Reinforcing Mistakes

Psychiatrists can use techniques to regress people into their childhoods with the intent of allowing them to express repressed emotion. The idea is to let the emotion out in the way that people were once bled to let out the vile humors. Sometimes bleeding worked by coincidence – lowering blood pressure may have had some effect on hypertension. Overall we know that bleeding was generally harmful.

Could this bleeding of emotion be the wrong way to go about helping people with emotional problems? It seems to me that it may be the lack of context in the original traumatic, or traumatic-seeming, event that causes a good deal of what people like to term as their issues[41].

If this is the case, regressing is just practising the same mistake over and over again until we have established a plastic rut in our brain. We are associating emotion over and over again with a painful memory – making it ever more painful and troublesome with each iteration. Each time we experience the emotion and associate it with the memory we are overloading the original event rather than releasing it.

Emotion cannot be treated like the pressure in a heating system. There is no store of emotion – there are associations and triggers to emotional responses. Change the triggers or the response and the problem often goes away. It is a problem of response not storage. Practising unhelpful behaviour does not change it, it reinforces it. By the same rule, practising helpful and healthy behaviour reinforces healthy and helpful responses.

Look at it this way. If you were to see someone pushing someone else off a railway bridge during a fight, you would be required to give evidence in court. In this court there is a time

---

41 Of course there are people who have been abused as children and otherwise damaged but that is not primarily what I am talking about here. However, the types of technique I am going to talk about can help even in those situations if used by people trained to deal with this type of trauma.

machine and you are obliged to go back along your own personal time line to witness the same thing again. You can arrive a bit early for the event, stay a bit later and you can stand anywhere you like.

If you had that opportunity and someone's life was in the balance of your understanding, would you go back to the exact moment and vantage point you held during the original event?

If it was not a push but an attempt at rescue and if the argument was not an argument but some clowning about, would the same vantage point help?

Going back and witnessing it from the same angle with the same lack of context would only serve to further convince you of your mistaken assumptions and misunderstandings.

Reliving old wounds is like taking a time machine back to the same flawed vantage point and cutting an even deeper wound[42].

## Practising Perfection

A more helpful method involves regressing, but not as a child. A child will still feel the total unfairness of the perceived situation. If you go back as an observer carrying the experience of being an adult and the ability to recognise the original event in context, it is an amazingly powerful experience.

---

42  This is one of the fundamental principles of NLP. Utilising strategies to refresh emotions and practice them is a bit like poking a sore tooth with your tongue to make sure it still hurts. NLP is often content free – it only uses the content of trauma to provide necessary context. NLP concerns itself more with how you react than what you are reacting to. Many practitioners consider delving into the details to be counter productive and that it offers people only an opportunity to practice the limiting belief or unhelpful neural state. The goal of NLP is often to help you practice more helpful states knowing that if you stop practising unhelpful ones that the brain will fill in the old ruts itself.

This is a useful principle to deal with any situation in which you feel you were hard done by, unfairly treated or one that you handled badly.

Once you are observing, you can imagine the resources that might help the people involved to gain a better outcome. In doing this you recognise the limitations of the people involved. Unhelpful blame and emotion can be dealt with much more easily.

Our real selves, when treated with this type of approach, often turn out to be rather nicer than we could possibly have hoped.

If we were doing our best and we misunderstand context, a callus can grow over the original event in layers. This distorts and makes us believe we have more faults than we have.

Like a thorn dissolving and the skin healing, many of the traits that we think of as shameful or dark and deep are just like infected matter around a thorn. The immune reaction is more painful than the thorn itself.

Understanding the context of the event as an adult or in the light of new learning allows us to remove the foreign body and let things heal. It is surprising how few deep dark secrets and vices we have when we shine the light of context on our memories and defuse the time bombs of confusion.

## Empathising With Others

Being yourself might be a lot more fun than you think. Those skeletons in the closet tend to evaporate into mist when you drag them out into the sunlight and realise that they are just illusions.

Allowing other people to be themselves, by admitting you may not understand the context of their behaviour, can be amazingly liberating. It is a great relief to be able to stop remembering why you are angry with someone.

When you replay some of the events in your personal history with this attitude or altitude you can consider what resources they were lacking. It fosters an understanding of the situation they were in at the time.

When looked at like this most situations are not as black and white as we often tell ourselves. When you can hear the other person's voice more clearly, because your inner child voice is stilled by the mature adult, things become easier to resolve.

# The Nature Of Consciousness

> *"My unconscious knows more about the consciousness of the psychologist than his consciousness knows about my unconscious."*
> **Karl Kraus**

## Interdependence Of Thought

When it comes to identity we are often tempted to label ourselves for convenience. There are many labels available and however plausible they sound they are all limiting.

Are you really left brained or right brained? Would your conscious mind really benefit if you were to suppress your unconscious mind? Would your intuition stop functioning if you examined the facts? Does analytical ability mean you have to avoid creative thinking for fear of losing your steely edge? Does a creative nature mean you cannot make tough decisions? Does habitual behaviour excuse you from deliberate action? Do reflexes diminish your self control?

Any exclusive identity choice means you lose something. All of the above are abilities and they are aspects of our functioning mind. They are as interdependent in our thinking process as legs are in our walking process. In both processes, balance and counterbalance are converted into motion. We move our consciousness from one to the other and perceive time and self in the transfer. Getting stuck in one state has serious consequences. These consequences are normally perceived as mental illness.

Much has been written about this and much more will be written, but I think it is possible to get a functional understanding without getting into all the semantics and fretting.

You have heard of precedent. In law if a court makes a ruling that can be recognised as a precedent, in the next case that is similar the court will use that precedent to reach a decision.

## The Internal Judiciary

Courts are hierarchical and precedents flow downwards. For instance, precedents set at a higher court are binding on courts lower in the hierarchy. Precedents set in lower courts may be overturned by higher courts. To challenge a precedent you must show that it is not relevant to the current or pending case. You must show that the significant facts of the precedent case are not present or that there are additional significant facts.

There are valid arguments in support of and opposing this type of system and any lawyer or solicitor will tell you that it is a bit more complex if you delve into the detail. I only want to remind you of the general principles here. I want to draw your attention to the similarity of this to the internal judiciary of the mind. That is the unconscious part of us that creates laws of self from precedents of behaviour.

## The Executive Mind

Think of the American political system[43] for example. There is executive power – the president who has the power to veto or sign laws into legislation. The president also has to execute and enforce the laws created by congress.

Most democratic political systems have similar ideas of the executive and the legislative being present. The executive tends to deal with the current situation while the legislative deals with some way of enshrining the long term implications.

For the purposes of getting at a working understanding of the brain we can imagine that the complex matter of

---

[43] http://www.whitehouse.gov/our-government/

consciousness can be thought of as having executive and legislative functions.

The executive side is your left brain and your conscious analytical mind. The legislative side is your right brain or unconscious creative side.

Logic emerges as the result of conscious steps; creativity appears as if by magic. Both logic and creativity benefit from practice.

Now imagine that the unconscious part of the brain is not really unconscious. Imagine it just has none of the trappings of power. It watches and records the actions of the conscious part until it thinks it has enough precedent to create a law. That frees the conscious part to rely on precedent and those that appear to work become deeply ingrained and respected.

This is a clever arrangement that allows us to function rather well. As we learn, we internalise certain things. Because of this we know what to do without figuring it out anew each time. This frees us to get busy learning new things. We can create a shorthand, or neatly compressed system, for dealing with the world. The unconscious mind feeds the conscious mind with precedent. The conscious mind is free to exercise executive power.

This is very important. It is perhaps the most important metaphor in this book. Through it we begin to glimpse what neuroscientists are trying to tell us.

## Learn To Relax And Love the Unconscious

Sigmund Freud introduced a horrible idea that the subconscious was a seething mass of neurosis and deep dark desires. This has had the effect of keeping psychoanalysts in shoe leather and the rest of us in terror of our own cognitive functions. Neuroscientists are confirming that what approaches, such as NLP and positive psychology, have been saying for years, may be closer to the truth. There is no seething mass of Oedipal, psycho-sexual, repressed aggression. Unless you want it, that is.

The unconscious is you. It can be either an advising angel or a tempting devil. It all really depends on what you tell it you want it to do. You listen very attentively to yourself and do your best to create the reality you think you want.

But let's get back to what you can do with this knowledge. How can you apply it?

# Arm Yourself With A *Priori* Knowledge

> *"We are all like the doctors*
> *We cut into life*
> *And we like to see blood*
> *On the end of our knife"*
> **Don McClean**

## Knowledge Is A Powerful Tool

History is full of people who fear knowledge. Maybe they see it as the highway to ruination. Maybe they seek to control it and contain it for their own purposes. Maybe they fear that knowledge without responsibility is dangerous.

The Greek myth of Pandora's Box warns that the search for knowledge culminates in the release of all evil upon humanity.

The bible asserts that self knowledge has lead to expulsion from paradise.

Mary Shelley's Frankenstein warns of the dangerous consequences of seeking the knowledge of the gods.

Knowledge misunderstood and used without consideration of the consequences can be dangerous and destructive. Knowledge is the ultimate tool. Any tool that is effective can be used to create or destroy.

Each experience changes us and creates a new version of us or of our model of the world. It destroys the old version. You might observe that it is almost as if we are reborn every instant and each following instant is an afterlife. It takes as much determination and skill to hold onto old models and ideas about ourselves as it does to adapt and grow.

## Philosophy Unbound

There are some writers, painters and poets who believe that art can only come about through the great suffering which burns away residues of self delusion. They appear to have concluded that everything except suffering and misery is self delusion.

There are psychiatrists, scientists and captains of industry who believe the same thing about progress. They believe that forward motion must be a struggle.

There are revolutionaries, governments and dictators who believe the same thing about social reform. They believe that blood and suffering is the price of freedom.

This can be perceived as subscribing to the philosophy of Hobbes[44] but if you think this you should re-examine your philosophers.

The philosophers are a fine body of individuals. You should listen to them courteously. They are learned and they are serious.

There is much in what they teach, but be careful about declaring allegiance to any one of them. Allegiance can set like concrete and restrict thinking.

They are learned but they are also jealous. Each of them tries to oust the others as charlatans. In doing so they can become

---

44 Thomas Hobbes (1588-1679) a political philosopher who proposed that the natural state of human beings was to be in perpetual struggle with each other. He argued that the only escape from this is to submit to the absolute authority of a sovereign or state and accept abuses of power as the price of a civilised society. He also subscribed to the notion of self interested cooperation. The character of the eternally stoical Tiger, Hobbes, from the cartoon strip Calvin and Hobbes is based in part on the philosophy of Thomas Hobbes.
http://www.philosophypages.com/ph/hobb.htm
http://en.wikipedia.org/wiki/Thomas_Hobbes
http://www.iep.utm.edu/hobmoral/
http://www.gocomics.com/calvinandhobbes

more interested in rhetoric and rationalisation than exploration.

We all have an inner philosopher who has already made up his or her mind about what is the word of truth. We have an inner gauge, to quote Ernest Hemingway, "a *built-in, shockproof, crap detector*". The challenge is to make sure we are not fiddling about with its crap detecting settings in order to allow the crap of our choice through. Even if this crap sounds academic and serious or is full of emotional blackmail, it must be detected as crap, whatever its source.

## Complex Equivalence

A lot of crap comes in its own Trojan horse of self reference, double bind, tautology and complex equivalence.

It might sound like this: "All good theories are complicated, this theory is complicated, therefore it must be a good theory."

Complex equivalence, self reference and double binds cannot be unravelled from within. They can be unravelled if you find a place to stand. Perhaps the best way is to stand under them and examine your position from this position of under standing.

This inner philosopher, *like all philosophers*, must be challenged and taught to make the distinction between wishful self indulgence and a full scale crap alert.

## Reality As A Metaphor

> There is a blade. It can cut reality. It is sharp. It can cut reality in at least two directions, possibly eleven. The knife can cut in any direction it is guided. It can cut toward dismay and horror. The knife can also cut toward joy and fulfilment. All of these things exist, in reality, for us.

The knife responds to the most subtle touch. It cuts deep into life.

It is the hand, mind and heart of the artist that determines the direction of the cut. The slicing of reality is a true thing to do.

Be careful of how you hold the knife though, it can cut the person who is cutting. Nothing is without consequences. Everything, every breath has consequences. The artist becomes her work. The artist is responsible for her work – there is no point blaming the muse.

# The Crucible

*"Suffering becomes beautiful
when anyone bears great calamities
with cheerfulness,
not through insensibility
but through greatness of mind."*
**Aristotle**

## Suffering

Suffering is a place. Once you have been there, you know its address. You know how it speaks and you know the only reply. I think that every person alive visits this crucible at least once.

If you have been there, it is not something you boast about, nor do you compare notes with others who have been subject to its tests. You do recognise their eyes though. Compassion is what you feel; and respect for yourself and them.

If you have sat by the bedside of a critically sick partner; if you have watched a parent die; if you have stared into the night and felt the helplessness of knowing that somewhere a loved one is in danger or in pain and that in the morning it will be too late; if you have heard the words that measure out the remaining days of your life; if you have heard your children cry for food; if you have walked the war torn streets of some city; if you have stared into the abyss of losing a child; or lost that child; if you have, then you have known the crucible.

It holds all fear and pain and once you have passed through it, nothing frightens you again.

I mention this because, even when embracing life, there needs to be the iron of knowing death in your arms.

I mention this because I need you to know that I do not dismiss the crucible.

I mention this because it is against this background that joy is possible.

I mention this because I want you to know that if you are in the crucible, I do not dismiss your pain. I respect your patience, your forbearance and your courage.

## The Anvil Of Life

There will always be times when we are beaten and shaped on the anvil of life. For all its terrible roaring silence, it is in the crucible that we glimpse infinity.

When even the seemingly unquenchable hunger of the crucible passes, what you will have discovered there will have opened your eyes, your heart and your humanity.

When dealing with people we must acknowledge that they have real lives in which real things happen to them. We must be aware that at any given time some of them will be enduring the crucible.

One needs to be careful to realise that out of pain, worry and fear, people can do the most unlikely things. They may not even realise what they are doing. When they do these things in your presence, they may be dealing with something happening somewhere else. They may be dealing with troubles you cannot imagine.

One of the most valuable things you can do, as a human being, is to stand back and look into the eyes of colleagues and friends and listen, not to their words, but to their meaning.

## Suffering Dislodges Certainties

There is no doubt that suffering often brings learning and maturity with it.

Why would that be?

Well the theory in this book so far is that learning happens when our model of the world is loosened. If you are unwilling to change your beliefs or your model, then you cannot afford to learn anything. If this is the case, you will simply search out things that conform to your model of the world and you will restrict yourself to opinions that conform to your own.

When exposed to great sorrow, tragedy, death and suffering, there are not many models of the world that remain intact. Everything seems to shift sideways and the familiar becomes uncertain. Suffering shatters many of the lenses through which we view life.

The same misconception surrounds the idea that good things must be earned or that the worth of something is proportional to the effort that went into earning it. Effort focus and concentration can indeed clear our mind of clutter and debris. It can make us much more receptive to progress and the sort of learning and insight that makes things happen.

What I challenge here are the complex equivalences at play.

Certainly suffering brings about learning and growth in many people but is the suffering necessary for learning and growth? Is there any learning and growth without suffering?

Of course there is.

## Learning Without Suffering

We seem to appreciate things more if there is hard work and effort involved. We can see what goes into them. There are two complex equivalences here that must be challenged.

The first complex equivalence is that **all** things that take a great deal of effort have worth. The second complex equivalence is that **only** things that take a great deal of effort have worth.

Many things take vast amounts of work, investments of money and sacrifice but are ultimately pretty worthless.

Many worthwhile and valuable things can come easily, effortlessly and immediately.

There is a certain clarity of mind associated with the state achieved through suffering or expending great effort. What if you could achieve the same clarity of mind with little or no effort?

Many people have a limiting belief that suffering, effort or pain is the pathway to success. They load pain and unnecessary complications into their life in order to invest value into something they want.

Sometimes suffering, time and drudge are unavoidable but they are not the inevitable price of learning, growth or worth.

Many of the greatest insights in the history of the world have happened in flashes.

Thomas Edison is often quoted as saying that every failure taught him something. This does not mean that he would not have been just as happy to find the right answer straight away. His discoveries and inventions would have had the same intrinsic value.

## Transforming Suffering Into Effortless Gain

As we discover how the brain works, it has become clear that we are constantly engaged in creating a reservoir of resource.

In many cases self imposed suffering or unthinking stoicism can impede the flow to and from this reservoir.

If suffering becomes a state, learning stops. When the mind is paralysed by despair and we acclimatise to suffering, it is tragedy beyond comprehension. When suffering has become the stuck state and when we recognise it, we ache for the person in this state. It is a confused tangle of complex equivalence and secondary gains.

Achievement can come about as a result of suffering and effort because suffering and effort can shatter our habitual models and show us the world in a new light. It can give us perspective. Anything that jolts us out of the rut can create insights and inspiration. Many people who undergo suffering can readjust their priorities because the suffering dwarfs the circumstances or events that were holding them back.

It misleading to think that it is the suffering and effort that are creating the insight and inspiration. They have merely led to the state of being unencumbered which is the wellspring of achievement. This state can be achieved without suffering and without immense effort once you have untangled the complex equivalence and exposed the secondary gain.

Do you remember any time in your life when you were really thinking clearly and things were just falling into place? Think about it now. Remember how things looked, sounded and felt. Think of how simple it all was. There you go!

We can use the metaphor of the channels created by the sled on the snowy hill. Suffering can jump you out of your channel so that you see the possibilities on the hill or it can itself become the rut.

The same can be said of working hard toward something. It can dig you out of your rut or it can dig you in deeper.

## Clarifying Equivalence And Re-Associating

Coincidence, secondary gain and complex equivalence are associated. Complex equivalence is when we connect an effect to a cause through circumstantial evidence or coincidence rather than through empirical proof (i.e. He loves me because he gave me an expensive birthday present.). Secondary gain is when we manipulate cause and effect to get the conclusions we think we want (i.e. if I give her an expensive birthday present she might tell her sister what a great guy I am.). Concealing secondary gain causes us to opportunistically create complex equivalences (i.e. of course I love you, didn't I give you an expensive present, now come to bed and don't tell your sister.). We deliberately, consciously or unconsciously, associate the cause with the effect to support our model of the world. The driver for this complex equivalence is almost always secondary gain and the opportunity is coincidence.

In terms of suffering and effort we want to justify what we have invested.

What actually happens is that pain, suffering, shock, surprise, novelty, inspiration and joy cause us to dissociate from our model. Struggle, striving, effort, work, practising skills and increasing our sensitivity to distinctions can challenge our model and cause us to dissociate from it and change it. (If you want to change the tyres on a car you have to stop it and get out of it – you can't keep driving along and hope the wheels will change themselves. If you want to buy a new car you have to get out of the traffic, get out of your car and examine the possibilities). If it causes us to dissociate long enough we can refresh our model for the better (learning) or we can completely abandon it for a radically better model closer to the deep structure of reality (enlightenment).

We create complex equivalences all the time with all sorts of things. The most pervasive of these, and the most clichéd[45], is cost and value. In fact we are so good at creating internally consistent models that we create compelling explanations when we really do not have a clue. If something works, we grab at the closest thing to hand that will provide a secondary gain and we give it the credit. In the crucible we credit the suffering, at work we credit the process or methodology, in life we credit selfishness and short-term-ism.

We can do the reverse equally efficiently with blame. We can blame our nature, nurture, parents or the gods for our suffering, we can blame the process, methodology or colleagues for lack of progress, we can blame our looks, our

---

45 "He/she knows the cost of everything and the value of nothing."

education or whatever else is at hand. The net effect is that when we create complex equivalences and hide the secondary gain, we can easily end up trying to fix the wrong problem or implementing the wrong solution. We can often find ourselves struggling with mountains when we could step over molehills.

The very abilities used to create complex equivalence and hide secondary gain are the ones required to clarify equivalence and re-associate with our true self interest.

## Dissociation And Models

In Section One I proposed that physical pain is a model that can be dissociated from. Professor Ramachandran has shown with his experiments that dissociating from old models and associating with new models through feedback can create remarkable results in people suffering from long term pain.

The same is true of mental and emotional pain. Gregory Bateson showed that schizophrenia can be both explained and treated with a concept of the double bind – a sort of extended complex equivalence with no resolution except dissociation from the current model.

The ability to associate and dissociate allows us all to deal with conflicting models within our own life. Models of work, life and play often cannot coexist in the same state. We all have the ability to move between associated and dissociated states. We associate with things we care about in the here and now and that we want to focus on. We dissociate in order to see the bigger picture and create choice and flexibility of behaviour. We learn and progress by dissociating from tired models of reality and seeking models closer to the deep structure that explain more and offer more potential for meaningful experience. We enjoy our life by associating into these models of reality until we are ready to move on.

## Professional Detachment

Let me take a practical example of another sort of complex equivalence. A surgeon's training teaches him or her how to detach from the emotional response of what they have to do. They have to cut into the body of another human being and sometimes make very difficult decisions. It is right that they should be able to be detached and professional at these times.

### *Short term gain long term detachment*

Some surgeons respond by deadening their empathy response altogether. They see patients as nothing more than a collection of symptoms.

These are the surgeons who declare that the operation was a success but the patient died.

> My father had a heart condition and he subsequently developed pains in his legs. A heart consultant recommended and embarked on a course of treatment for this symptom. The new medication made my father very ill so he went to see his local doctor. His doctor told him the medication he was taking for his legs was very dangerous for someone with a heart condition. When the consultant was asked about this he agreed that the medication could kill my father. He insisted however that the treatment was the right one for the leg pains.

### *Long term gain short term detachment*

There are other surgeons who suspend short term empathy in favour of long term empathy. They know that they must be objective in order to make difficult decisions and to perform complex and invasive surgery. Their advantage is that they have more than one state of being, as a surgeon. People like this realise that suspending empathy altogether is as dangerous to their patients as being too emotionally involved.

The ability to detach or dissociate is valuable and necessary. It can also become a rut.

## Objective Associations

It is the contrast between being associated and dissociated that gives us perspective. It is the movement between the states that teaches us where we really are. Many NLP techniques use this deep principle in order to help people change their models of the world, shake limiting beliefs and overcome phobias and emotional blocks.

## Avoiding The Crucible

It is said that we only hurt the ones we love. It may be equally true that we can only be truly hurt by whom and what we care about. Anyone who has children knows that they have taken hostages to fortune into their heart. The capacity for joy and pain increase together.

To care about something is to associate. Associating is a process of identifying.

If it is something whose outcome we have no voice in or whose outcomes we find in some way distasteful, it can quickly become a crucible of sorts.

If it is something in which we perceive unavoidable dependencies – like our job – our choice seems to be either to associate with something whose outcome is likely to bring pain (and accept the pain as a necessary cost) or to dissociate from it and to have a meaningless waking life (and accept the suffering as a necessary cost). The *either-or* choice should be an alarm that the situation is fraught with complex equivalences and false associations.

The verb to earn, to many people, means to pay a high price in terms of suffering, effort or self denial. There is nothing wrong with being prepared to work hard or to endure to achieve something. It may be a legitimate way to estimate cost but it is a crazy way to measure value.

Here is something to consider when we think about work:

Page 90 | *Section Two: The Nature of Knowledge*

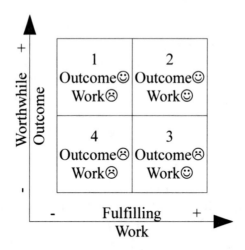

1. You are working toward an outcome you want but you hate the work.

2. You are working toward an outcome you want and you are enjoying the work.

3. You are working toward an outcome you don't want but you are enjoying the work.

4. You are working toward an outcome you don't want and you hate the work.

Keeping this set of quadrants in mind, let's talk about the intrinsic value of work and the concept of delayed gratification.

## Delayed Gratification And Selfishness

Many methodologists and financial and economic gurus argue that people work in their own self interest. They make process and policy based on this assumption.

Most of us work in our short, or at best, medium term interest. If that happens to coincide with our long term interest, that's fine with us, but in general our long term interest is of no interest to many of us.

We see our future self as a stranger who can look out for him or herself. We all take decisions that will make things very difficult for our future selves but that will give us immediate gratification.

We do not need to look far for everyday behaviour that demonstrates this –

- Habitual overeating:
    - present self gratified
    - future self overweight
- TV couch potato:
    - present self gratified
    - future self unhealthy and dumbed down
- Smoking:
    - present self gratified,
    - future self lung cancer
- Habitual alcohol abuse
    - present self gratified,
    - tomorrow's self hung over,
    - future self a raft of health problems
- Borrowing to live beyond means:
    - present self gratified with shiny things,
    - future self in debt

- Wasting non renewable energy
  - present self gratified
  - future self will miss the polar bears and fresh air

Our ability to delay gratification is seen as a mark of maturity and civilisation. One of the ways we chart the development of babies is to measure their ability to delay gratification.

## Intrinsic Value

> I was faced with the prospect of pointing a stone wall. On top of it being a hot summer that year, I had several long term conundrums and worries to sort out. I had some very serious decisions that had to be made.
>
> I looked for short-cuts but there were none to be found. The limestone render had to be pushed in methodically with a small trowel between stones that ranged from the size of my fist to the size of my head. It looked like a boring job and it was going to take weeks.
>
> I decided to look for intrinsic value in the job. Doing it well would present me with a beautiful end wall to my house that would resist the weather for the rest of my life. So I set up my scaffolding, cleaned the wall and mixed the first batch of limestone mortar.
>
> As the job progressed I began to get more mortar between the stones than on the ground. I could feel the confidence in my hands growing and there was pleasure in it.
>
> The sun was shining on my back and my hands were almost working on their own.
>
> Each evening I came down from the scaffolding tired but relaxed and easy in my mind. I stood and admired my work and found that I was thinking my way through other decisions I had to make as the work progressed. I had decided that as the wall was finished so I would arrive at the decisions I needed to make. My mind kept its part of the bargain.

> On the day the wall was completed I was proud of my achievement but I was sorry the work was finished. I had been enjoying the work and the restful and thoughtful state it had triggered in me every day.
>
> The work had intrinsic value to me. It was valuable in itself. The doing of it was satisfying minute by minute, hour by hour, day by day.

People do their best work when it has intrinsic value to them. I am lucky to have worked with a great number of programmers for whom programming contains a huge intrinsic value. Some of them care about the end result but many care only about the quality of their code and what it says about them as professionals and clever people.

## Money Is Not A Motivator, It Is A Demotivator

When Frederic Herzberg[46] looked into motivation he found that salary was not a motivator. In a strangely counter intuitive twist, it appears that we are demotivated when we are not being paid enough but that we are not motivated by more money. There are also clear correlations between overpayment and the lack of performance.

He also found that the intrinsic value of the work itself comes very high on the list of things that motivate people.

The motivators he found (in order of potency)

- Achievement
- Recognition
- The work itself

---

46 Frederick Herzberg (1923-2000) – Psychologist. Most famous for his work on *job enrichment* and *motivator hygiene theory*. His books include "the motivation to work" (1959) and *"work and the nature of man"* (1966). His theories simply stated were that "the factors which motivate people at work are different to and not simply the opposite of the factors which cause dissatisfaction".

- Responsibility
- Acknowledgement (promotion)
- Growth

The demotivators he found (in order of potency)

- Company policy and administration
- Supervision
- Relationship with boss
- Conditions
- Salary
- Relationship with peers
- Personal life
- Relationship with subordinates
- Status
- Security

## Balance, Counterbalance And Pay

In volume one, we talked about balances and counterbalances in managing organisations. Counterbalances are quite often not opposites. Not paying people enough will demotivate them but paying them more will not necessarily motivate them. Not being demotivated is not the same as being motivated.

### *Working for intrinsic value*

People will sometimes work for just the intrinsic value of doing something. Consider Wikipedia, Sourceforge and other collaborative, non profit ventures. Many people volunteer their services to good causes and often this is just because they feel it is the right thing to do. Is there anything better than coming in from the garden exhausted or the feeling of having fixed something yourself even though it would have been inexpensive to get someone else to do it?

### Low salary perceived as exploitation

The fact that money is not a motivator is not an invitation to force people into lower salaries. People who are not paid properly for their work will be demotivated, even if they had previously been willing to do it for free.

The minute people feel that they are being exploited they can abandon even the most lofty ventures. I would bet that the reason that Wikipedia attracts such cooperation is that Mr Wales is not making a fortune off the back of it. It might be tempting for someone to think that it would not make any difference to the people doing the work even if he were making money. Not so. People just do not like the idea that their work has been taken for granted. By contrast to Wikipedia look at the furore[47] over the sale of the Huffington Post.

### Reward as licence to free-load

Counter intuitively, people who are paid too much for their work will be equally difficult to motivate. Recent studies show that productivity drops off as wages rise above a certain level. That level is the level that takes money off the table as an issue. When you start to try to motivate people with money as a reward, they stop performing.

In his book, "Drive", Dan Pink outlines the science of motivation and examines recent studies and research that demonstrate that people are not as easy to manipulate as many management theories would have us believe. When tasks require even rudimentary cognitive skill, rewards do not

---

47 http://www.guardian.co.uk/media/2011/apr/12/arianna-huffington-post-sale

improve performance, they damage performance. For simple straightforward tasks (algorithmic work) carrot and stick works very well, but as soon as you involve complication and conceptual thinking (heuristic work) this method starts to demotivate.

There are several theories as to why over-payment adversely affects motivation, but my own is that people start to feel an over-inflated sense of the worth of their time. "Why should I do that, I am getting paid X, that task is beneath me and I don't need to try." I think that people need to feel a connection between the intrinsic worth of the work they are doing and what they are being paid. If not they feel they are getting away with free-loading and don't see why they should work.

### What is intrinsic value

People are motivated when their work has intrinsic value to them. That means that they want to feel that there is some point spending their time doing it. People like to feel that their contribution makes a difference, but intrinsic value is something slightly more than this.

Let's take a closer look at intrinsic value. Money has very little intrinsic value. A banana has more intrinsic value than a £20 or $20 note. The banknote has extrinsic value. The note is all about perception. You and the people around you agree that you will all swap goods and services for that specific piece of paper. If you could take the note back in time or in some other way get it out of the shared model that gives it extrinsic value, it is practically worthless.

The banana has intrinsic value. Regardless of whether you like bananas or not, they continue to contain high levels of vitamins including high levels of B and C. They also have high levels of potassium which has a protective effect against hypertension and magnesium which gives an instant energy boost. If you rode your bike to work this morning, eating a banana will do you lot more good than eating a banknote.

One sure way of giving intrinsic value to work for knowledge workers is to ensure that the work is constantly nourishing and increasing their knowledge and experience. On an existential level, we know intuitively that the most intrinsic value we can glean from something is when it increases our own intrinsic value by increasing our reservoir of conscious and unconscious resources.

# The Reservoir

*"Music is the pleasure the human mind experiences from counting without being aware that it is counting."*
                                                    Gottfried Leibniz

## Connecting To Experience And Knowledge

The more experience and knowledge you have, the more there is for intuition to work with. Intuition works through connecting all your conscious and unconscious experience and knowledge.

In NLP and cybernetics this is sometimes called transderivational searching. It is a joining of models with potential or alternative models and searching through your unconscious reservoir of knowledge and experience.

### *Computing*

In computing it is a fuzzy search using content addressable memory[48] and associative arrays[49]. This means that when you search on a value, you get a list of the places where it occurs regardless of context. Fuzzy logic is only a hop, skip and a jump away. Fuzzy logic deals with approximations rather than concepts of true or false. It is used in artificial intelligence as it is a closer match to human intelligence than binary logic.

---

48 Content Addressable Memory (CAM) is a type of memory used in high speed searches. Normally memory returns the physical addresses (the physical location) of the data and the data is retrieved from that address (location). CAM searches the whole memory for instances of the data term and returns a list of addresses where it was found. This resembles an inverted index. Think of the way an index normally works and invert it. Normally you access data through indexes which point at some array of data. Invert it and the each datum has a list of its own occurrences.

49 Associative arrays are arrays of data where there is a one to one mapping between the key and the value. Keys will have only one value. Values will have only one key. Two or more keys may have the same value. E.g. Key(Name) Value (April), Key(Month) Value(April). But you will not find a year with two months called April.

## *Remembering*

How do you find anything in your mind? If I ask your age; your mother's maiden name; your fathers middle initial; the colour of your first car – how did you find that information?

We just seem to know. Even when we have not been thinking about it and especially before we start to think about it, it just arrives into our conscious mind. It is as if something just handed you the answer. You call it memory. It was not a conscious mechanism that knew where those memories were stored. We have a silent partner on board. Our unconscious mind carries out one of these transderivational searches through googolplexes[50] of data.

## *Talking*

When we express something in language, we approximate thoughts with words. This requires a mini transderivational search. Our unconscious supplies the approximate words.

How often have you felt that you could just not get the right words for your thoughts? Normally this happens when we are trying to express complex emotions. However often, it was insignificant compared to the number of times you opened your mouth and the words magically appeared.

---

50 A very big number – 10 to the power of googol. Googol is 10 to the power of a hundred. Googolplex is 10 to the power of 1 followed by a hundred zeros. Trying to express the memory capacity of the brain in terms of computer storage is fraught with difficulty, not least because of the glaring presupposition, required to compare it to the computer model, that it stores memories digitally. Some scientists, including Jonathan Von Neumann, have suggested a number of between $10^9$ and $10^{20}$. There are 100 billion neurons in the brain with the possibility of around 1,000 connections each. If we consider that memories are made up of configurations of neurons, that this is probably not a digital process, that there are more possible combinations for this than particles in the universe and that there may be a lot of quantum level activity involved, then googolplexes make about as much sense as any other estimate and it sounds a lot more scientific than "shitloads".

## Reading

Just think about what you are doing while you read this. Where are the meanings of the words coming from?

## Thinking

When we think, where do the thoughts come from? Thinking is associational. Something reminds you of something that reminds you to think about something else. You hear a tune and it sticks in your mind. You hear a joke and it reminds you of another one. And that reminds me....of all the things I know.

Each of these processes starts with a transderivational search. The nature of learning relies on transderivational searching and association which are more like fuzzy logic than the razor sharp logic machines we sometimes imagine we have in our head. Scientists have been trying to create this sort of computer intelligence since something first sparked off the idea of computers in the first place. How many ideas are sparked? What does a spark do?

All thinking, talking and acting requires us to interact with our unconscious mind. Each interaction requires something like a micro hypnotic trance as we create logic out of the most illogical cloud of associations.

Did you enjoy your last meal?

Answering that required you do a transderivational search, both to sort out the ambiguity and to recall your most recent meal, then to sort out what memories you have associated with it. All in all, it is a pretty amazing feat. Computer programmers go green with envy at the thought of the elegance and the speed.

## *Intuiting*

So if we have to communicate with our unconscious mind to talk, to reason and to remember, is it such a big leap to understand that intuition is the product of experience, and that it can be practised and perfected? It is a matter of building a relationship with our silent partner. Earlier it was suggested that the beginning of everything was to empathise with ourself. This is literally what we have to do. Acknowledge and accept the nature of our own consciousness. Embrace the little bit of chaos theory in our cognitive process and the phenomenal ability that thinking is. We have to loosen the limiting beliefs we have constrained our thinking with and learn to listen to the voice that gives us answers out of apparent thin air. It is what we do best.

## The Quality Of knowledge Is Filtered

Our silent partner is a tireless worker. Like all workers, the more responsibility and trust you give it, the more it will take on for you.

It follows the basic rule of plasticity, the rule of action and reaction, the force of habit and the principle of feedback that applies equally to motivation, inspiration, appreciation, encouragement and meaning.

The better the information it has to work with, the more effective this worker is. For instance if you ask your intuition for a solution to a dilemma in physics and mathematics but you have never opened a book about them or been even vaguely interested in these subjects, it will do its best, but it will be working with all the impressions, limited understandings, lack of attention and misapprehensions that you have fed it on the subjects.

It is sometimes hinted that Einstein was the great outsider that changed physics. You must remember that although Einstein may have been unconventional, he was interested in physics since he had been a young boy. He may not have been recognised as a great student; but that means less and less as we start to understand how limited our education and academic processes really are when compared to what we know about how the brain works.

Einstein worked as a clerk in a patent office. The temptation is to think that this made him an ill educated outsider who came to his work as some sort of savant. Apparently he loved the job because it gave him plenty of time to read and because he was surrounded by the patents which represented the creativity of others.

His phenomenal theory of relativity did not come out of the ether. It came as his brain put together all the things he read and studied. He applied his conscious brain to asking questions. It is not at all surprising that his unconscious threw up idea after idea to his conscious mind. He respected his mind and fed it with quality knowledge.

The other thing to bear in mind is that Einstein loved thinking. In reading his accounts, it is clear that working on these problems was pure joy for him.

# The Shape Of Knowledge

*"Physics is mathematical not because we know so much about the physical world, but because we know so little; it is only its mathematical properties that we can discover."*
**Bertrand Russell**

## Mathematics, Logic And Philosophy

Plato said the highest form of pure thought is mathematics. Galileo said that mathematics was the language in which the universe was written. Einstein said that mathematics was the poetry of logical ideas.

Mathematics is the expression of logic and knowledge. It is the language of reason. It is the closest we can get to certainty. It is the extrapolation of the evidence of our eyes and ears that 1+1=2.

Yet mathematics has a hole at the centre. Maths rests on many axioms that have never been logically proven. For instance, from this perspective, 1+1=2 is an axiom based on anecdotal evidence.

When Bertrand Russell and Alfred North Whitehead set out to close this hole they were sure that they could do it. Maths is about counting and everything can be counted. Maths is about logic and logic is... logical. If there are inconsistencies it was just a matter of getting the proofs right, wasn't it? It took them hundreds of pages but they did prove that 1+1=2. This was only one axiom from many.

George Cantor had proposed set theory which allowed them to step up to a new level of using logic to prove mathematics. Sets appeared to tame infinity

Yet Russell himself proposed the paradox that almost destroyed his own work. Does the set of things that do not contain themselves, contain itself? If it does it doesn't, if it doesn't it does.

Gödel followed that with the hammer blow of the incompleteness theorems. He demonstrated mathematically that mathematics cannot prove itself true. In the minds of mathematicians there was chaos and logic was pronounced dead[51].

It would take a lifetime of discipline and a certain sort of mind to comprehend the mathematics of all this and, at first glance, it may not seem immediately relevant or important.

Any yet it is important. It is very important.

Let me explain how it all fits.

## Infinite Paradox Machine

In volume one I made much of Archimedes and his circles. I constantly referred back to circles in every metaphor I could plausibly roll in.

Circles are infinities. Basic maths tells you that a circle must contain an infinity of points in its circumference because to be a circle it must express an infinity of angles.

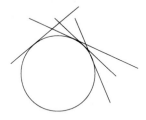

A circle defines infinity. It has no beginning and no end. Nothing is bigger than infinity. That is the point of infinity.

---

51 Jonathan Von Neumann, who was there, said "It is all over." Von Neumann was a renowned mathematician who contributed to quantum mechanics, set theory, game theory and was responsible for the Von Neumann Architecture, which was the idea of stored programs and upon which all modern computers rely.

On the circle there is a mathematical infinity of points. There can be no greater number of points possible than there are on this circle, if we are to believe in our interpretation of infinity. Nothing is bigger than infinity.

Imagine you could land on that circle and observe points infinitesimally close to one another. Imagine a line from each point crossing at the centre of the circle.

Now imagine that there is another circle. It has the same centre as our original circle but it is bigger. Imagine our two lines extending to this bigger circle. What do you notice?

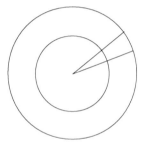

No matter how infinitesimally close the points are on our original circle there will be more space between them on the outer circle and therefore more points. The two infinities represented by the circles are not equal.

We have discovered different infinities and that they are measurably different sizes.

The circles can go on getting infinitely bigger or infinitely smaller meaning that there is an infinity of infinities.

Imagine the circles are being drawn on a huge sheet of paper. Imagine you can look at it from the side even though, if it is a mathematical plane, it is infinitely thin.

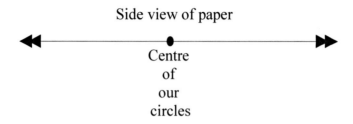

Imagine you can turn that sheet of paper around the centre. Perhaps like a board on a hinged bar or a table-football player on its rod.

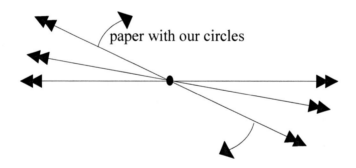

Obviously you will rotate it in a circle and we already know that there are in infinite number of angles on the circumference of the circle. There will be an infinite number of sheets of very thin paper thus making it an infinity of infinities of infinities of infinities.

That is just one plane of circles rotated through one position of our rotation rod. Now imagine that rod has a ball joint at the centre capable of rotating in an infinity of angles. Imagine all the points on all the planes as ball joints. Even with this we have not touched the possible ways of counting the different infinities.

## Russell's Paradox Revisited

Why does the pragmatist in all of us rear up at this sort of argument? I think it is because each of these infinities is a model. Maths is a model of reality made up of an infinity of models.

Maths is itself one of those circles. It is a set of things that does not contain itself. When you try to contain it within itself, it becomes a paradox.

If mathematics, the bedrock of all we hold dear in a rational world, is this... well, weird, where does that leave us?

Well it does not take away maths. Mathematics and logic are still the most powerful tools that we have. Because you cannot use a hammer to hammer itself does not make the hammer any less useful at hammering nails.

If you read volume one you will remember my Russian dolls within boxes within onions and all the ways we talked about recursion and self similarity.

Everything you do, every tool you choose, every decision you make, every perspective, belief, value or thought you have is one of those circles.

Try to use it to examine itself and it will throw you, unknowingly, into a recursive loop. You can call it paradox, you can call it cognitive dissonance, you can call it prejudice, but if you want to gain wisdom, you must jump to the circle outside your working level to understand principles, rules and proof.

Anything else is just matching or mismatching details on the map.

Imagine the circles are concentric spheres. Think of them as energy shells around a nucleus. Electrons jump from one energy shell to another by gaining or losing energy[52].

In the infinity of points that you can be orbiting – and we orbit many simultaneously – there are levels or shells of perspective.

## Energy Horses

When we jump energy levels in one orbit, all of our other orbits are affected.

If you gain understanding in anything, cookery for example (and I mean a real insight here), this generates understandings and learnings in your work, your hobbies, your relationships and your appreciation of almost everything else you are actively involved in.

It is a sort of auto synchronisation. We seem to be simultaneously co-central at many points of interest.

---

52 In 1900 Max Planck announced $E=h\nu$. Radiation (e.g. light) is emitted, transmitted or absorbed in energy packets – quanta- determined by the frequency of the radiation (v) and plancks constant (h). This pointed the way to quantum theory and Planck got the Nobel prize in physics in 1918.
*Interesting note: Planck was one of the first people to immediately recognise the significance of Einstein's theory of relativity and used his influence to have it accepted and was instrumental in advancing it. He did not at first appreciate the significance of his own remarkable achievement and it was Einstein who was among the first to understand its deep significance.*
In 1913 Neils Bohr proposed that electrons orbit the nucleus in allowed orbits, allowed energy levels or energy shells. While they are in these shells they do not emit radiation. They change energy shells by emitting or absorbing energy in quanta. The idea of orbits has been somewhat superseded but the basic idea of energy shells is intact. The further out the energy shells are, the more electrons they can accommodate. This is the chemistry and physics of matter - physical reality.

Cinema and the history of motion picture can be fascinatingly insightful. There is a film about film which shows one of the first moving pictures. It is a sequence of a horse jumping a fence[53].

Looking at it as a moving picture, the horse gracefully leaps over the fence. Looking at it as a series of pictures, the horse seems to move forward through a sequence of possibilities in space and time. The horse only inhabits each possibility for an infinitely small moment before the essential focus of the horse moves on.

## On Newton's Beach

I think therefore I am.

The world is as it is and it can only be thus.

Quantum physics talks about collapsing wave fronts. It is talking about the world of the very small and it is never very practical to try to make it apply to the world we can see, hear and taste.

Nevertheless it is tempting to imagine the horse's consciousness taking its route through the myriad possibilities that exist. As it goes you can almost hear the

---

53 Eadweard Muybridge was a pioneer of motion photography using multiple cameras to capture images of animals in motion. He also figured out how to slow down motion that was too fast for our eyes and freeze moments in time. Many of his films appear to me to be strange loops. They end at the start and look like continuous fluid motion when played in a loop.

crashing waves of probability as they collapse into motion and presence.

Flying outward on times arrow we collapse wave-fronts of probability as we hurtle along our trajectory. Each decision collapses the waves in front of us, allowing reality to set in and guide our flight.

This may seem a little far-removed from project management and cooking dinner, but it is essential to our understanding of what comes next.

With all this infinity lying around us like the pebbles on Newton's beach, we still have to breath, eat and attend to our human needs.

The decisions we make have consequences. None of them are inevitable, but some of them make others more probable.

Reality is like this set of infinite concentric circles. The process at your job is one. Trying to use the process to change the process is not possible. The process imposes limits and constraints as stringent as those of mathematics.

Trying to solve any problem requires you to move to the next energy shell. You must, like Archimedes when he said that he could move the world with a sufficiently big lever and somewhere to stand, find somewhere to stand.

## Deal Or No Deal

This is a game-show in which a contestant chooses a box containing some unknown prize. As other boxes are opened or discarded, the probability of what is in the contestant's box changes. They are repeatedly asked to swap this box for a known prize or cash incentive (deal), or to stick with the box (no deal).

Mathematicians and logicians want, more than anything else, to be certain. This is impossible. Each certainty in each circle is built on axioms from the circle containing it or orthogonal to it.

No matter how much you know, no matter how far you travel, you will always be basing your reality on axioms which cannot be proven in that reality. When you move to a reality or a set where they can be proved, you find you are resting on a new set of axioms.

So we choose realities, processes, methodologies that seem to fit the task in hand and get on with it. This is fine as long as you know that there are outer circles, intersecting circles, orthogonal circles and an infinity of circles to which your current circle appears to be the height of madness.

When you know this you can navigate as appropriate.

Although this word "appropriate" appears not to be very concrete, it can be measured. Reality is random. Organisational processes are random so that as their details are examined they present an impenetrable façade. If we move to an outer circle, suddenly there is enough room to drive a bus through what seemed watertight just moments before. Vague appropriateness in the last shell is now patently obvious as a quantitative measure from the new perspective.

You always have more choice than you think you have. Tools such as Agile and NLP are useful when they force you to change perspective. This puts you in a better position to know whether it is deal or no deal time.

Before you make a deal or no deal, you must realise that the box has no intrinsic value, it has the extrinsic value of what we believe it might contain. Logic and probability help us deal with the risk when we have gained perspective.

## Getting Involved

When Wilbur Wright was working on inventing the first practical flying machine he commented in a speech to the

Western Society of Engineers in 1901 that there are two ways to learn to ride a fractious horse. One is to get on and learn how each motion and trick can be best met and the other is watch, take notes and retire to the house to figure out how to overcome its jumps and kicks. The latter, he said, was safer but the former produces a larger proportion of better riders.

As we start to picture reality as a set of enclosed spheres or energy levels, we realise that we can traverse them in a variety of ways. We can orbit a particular context or model. We can travel outwards through enclosing realities. We can travel into the reality of a reality.

There is always a difference between watching and doing. Good coaches know this when they stand at the sidelines and watch the athlete. Even as they live the motion, they have to draw back and observe from the stillness of the next energy level. As the athlete internalises and feels the pull of gravity on the muscles, the coach must analyse the trajectory and the arc. The coach knows that this is not enough, that all of this must be fed back and linked to the internal experience of the doer.

Doing is a great teacher.

> The ski instructor watched me turning. I was struggling with the transfer of weight from one ski to the other, required to execute the turn smoothly. The result was a strange two step motion. He could have explained, as had all the other instructors, about how the skis worked and he could have demonstrated yet again how the turn should look. This tall, thin man was 68 and had undergone a complete liver transplant the year before. He skied as if he were floating an inch over the snow. He wasted nothing. He was a personification of elegance. Talking to him, I felt he could read my mind. He joined me, looked down the hill and said "Do the turn again. Forget about everything else. When I tell you to, grasp a dime with the toes of your downhill foot." When he called on the next turn, I did just that. I shall never

> forget the magic of that turn. A windhover, I swept clean on that bow bend. He had managed to dismiss my model of the turn. He had interrupted my strategy for failing. The doing of it gave me a new model of how to do it. He had known what was going on inside me as well as outside on the skis.

## Conclusion

Now listen closely. This is important. If you have been thinking about the dive we have taken, we are about to hit the surface of the sphere and carry on through.

Models are a tool. When we want to measure the immeasurable, move the unmovable and reverse time's arrow itself, we build a model and we bind immensity, eternity and infinity within its walls. We are so powerful that we capture the abstract and bind the intrinsic to the extrinsic. We forge a complex equivalence and we use it to shape reality. We hammer that piton solidly and, confidently, we start to climb.

We have such an ability to believe, that we create realities for ourselves. When we need to we can believe that the inside of the box is the sky and that the shadows on the wall are real.

We can bind value to pieces of paper and pass it around. We can bind value to electrons and put the world on a piece of sand. We believe so much that we forget that the banknote is just a promise and that the chip is a binary dream. As managers, coaches and teachers, we have to deal with the crisis when the coupling inevitably comes loose or breaks.

People can believe so hard that they confuse the intrinsic and the extrinsic. The extrinsic that has worked lose from the intrinsic becomes the goal.

Billionaires, who have forgotten what money represents, gather more and more of it. They buy power and use the power to buy more money. We have all seen the disaster films where some fool dooms himself by staying behind with the treasure in the collapsing caves. Life is a collapsing cave.

Money is great, it is a fabulous idea. Process is great, it is a magnificent idea. Democracy is great, it is a powerful idea. Business is great, it is a useful idea. Philosophy, science, mathematics are great, they are infinite ideas. Ideas are models of the world, ways to explain and navigate.

These are models that allow us to function. When any model becomes too rigid, it stops us from accessing the intrinsic value it allowed us to control. How good are we at creating models? We are so good we forget that they are models. When we do that, we hold onto the banana skin and throw away the banana.

But if we are that good, we can let go of any model because we can build another whenever and wherever we want. When we let go of models, we start to see the intrinsic value again. Models are fine, but we must always remember that they are models.

Boundaries are not constraints. Boundaries create a space for us to work.

When it comes to doing things, it all boils down to this:

## Summary:

We create models in the form of processes, rules and boundaries so that we have the means to achieve intrinsic value. When we look at them from the outside they allow us to handle the space and time they enclose. They are tools not masters. This is the purpose of processes, strategies and models: to bind the intrinsic and abstract to the extrinsic and concrete.

When we use control to create boundaries, we must prevent them from becoming constraints. Control is skill and leadership is remembering how it looks from both the next energy level and the one occupied by the doers. It needs the double vision of the shape without and the detail within. A leader is the fulcrum between the big picture and the detail and must respect both.

People need a mix of security and freedom. They thrive on a mix of autonomy and purpose. The context in which you present the working space and your ability to rearrange the boundaries can make it a prison or a playground of innovation and mastery.

# Managing Knowledge

*"Everyone has their nose in the rule book as if it was supposed to be an instruction manual."*
Philip K Howard

## The Information Age Effect

The world is a noisy place. Humanity is gathering knowledge and information at an unprecedented rate. Each of us is bombarded daily, at home and at work, with a deluge of information.

We know that much of what we need to know is hidden in plain sight. We have experts giving us diametrically opposing advice in everything from diet to global warming. We have opinions, facts and figures streaming into our life through computers, television, radios, ipods, books, magazines and endless emails, tweets, social networking sites and blogs.

Working and living in the information age, we have to contend with an education system designed for the industrial revolution. We are expected to get things done, despite processes that reward productivity over intelligence, quantity over quality and speed over suitability.

Managers, teachers, doctors, politicians and markets use combative, adversarial and competitive tactics to push the will of powerful interest groups over the concerns of the majority or of the individual.

Rules and processes seem to have taken precedence over consequence and circumstance. It sometimes appears that the breaking or enforcing the rules has become the main issue in place of understanding or examining the sanity or the intent of them. Who decided that presentation was more important than content; appearance more relevant than intrinsic value and that by mapping surface structures you could ignore the deeper structures of life?

### The hiding response

Is it any wonder that many people find themselves parroting sound-bites and half understood arguments? It sometimes seems that trying to talk rationally about things is like trying to catch sheets in the wind with your teeth, whilst blindfolded on a trampoline.

Many seek solace in the familiar. They wonder why everyone keeps telling them to embrace change, when the mandated change seems to be at odds with their own interests.

### The drowning response

Have you ever found, that you had everything you needed to avoid the crisis but that you could not see the woods for the trees? Was it a case that the necessary information had come wrapped in so much other superfluous information, that there was no way on earth to find it? Have you ever wondered why your job involves processing a tidal wave of information that leaves you no time to act on anything important you might find?

You are not alone. Almost everyone I talk to, in every profession, has expressed similar thoughts.

The best place to hide a book is in a library. The best place to lose knowledge is to drop it in the sea of information.

### The adversarial response

The world has become so complex and so adversarial that we have been conditioned to demand everything here and now. We do this for fear of missing something or missing out on something. In consequence there is a tendency to leap on the slightest mistake, ignoring everything else. By focussing on only the flaw there is a risk of dismissing, and thereby losing, the benefit of much that was done right.

### The hoarding response

Fear of losing out is a phobia. It causes us to tighten when we should loosen our grip. It causes us to close our fist like a monkey caught in a monkey trap. This, it seems to me, causes many of us to miss our very lives. We have forgotten to question why we do things and we only ask how. The telescope seems to be turned the wrong way around.

### The requirements response

In software projects, the stakeholders have been conditioned to ask for everything they could conceivably need. There is often no thought of how this behaviour might be degrading the quality of what they get by distracting the programmers from understanding what their customers need and why.

### Far away fields are greener

All things are not equal. Some theories are based on belief while others are based on evidence.

All requirements are not equal. Mostly the shiny attractive ones are worth less than the utilitarian pillars they rely on.

It sometimes appears that we are obsessed with cost and impervious to value. I wonder if we covet what we might be missing at the cost of what we have gained.

> There is an old fable about a dog crossing a stream with a bone in his mouth. He sees his reflection in the water and thinks it is another dog. He opens his mouth to snatch the other dog's bone and loses the one he was carrying.

Our supermarkets stock beautiful looking but tasteless fruit and vegetables. The most mundane things are packaged as marvels. The marvellous is sold as mundane. We demand perfection and accept mass produced rubbish that we have been conditioned to desire.

### *But the roses smell great here*

It is also a wonderful time to be alive if you are determined to be alive.

You can sulk that the flight was twenty minutes late taking off or you can sit in an armchair at 30,000 feet and wonder at the miracle of flight[54].

At the touch of a few keys you can explore the great literature, art and music of the world. We can explore the museums, art galleries and natural wonders from our own homes if we wish. We can talk to each other from opposite sides of the planet. We have the technology to take much of the drudgery out of work and create lifestyles for ourselves that encourage and fulfil.

Why is it then that so many people still live lives of quiet desperation and go to the grave unsung?

> "Men and boys are learning all kinds of trades but how to make *men* of themselves. They learn to make houses; but they are not so well housed, they are not so contented in their houses, as the woodchucks in their holes. What is the use of a house if you haven't got a tolerable planet to put it on? — If you cannot tolerate the planet that it is on? Grade the ground first. If a man believes and expects great things of himself, it makes no odds where you put him, or what you show him ... he will be surrounded by grandeur. He is in the condition of a healthy and hungry man, who says to himself, — How sweet this crust is!" – Henry David Thoreau in a letter he wrote in 1860

This next section proposes three more simple ideas that build on the three simple ideas from volume one[55].

---

54 Look up Louis CK 'Everything's amazing and nobody's happy': He is talking about a guy who can't get broadband on his flight and how quickly the world owes him something he only knew existed 10 seconds ago.
55 Context, balance and inspiration were the ideas in volume one.

# Three Simple Ideas

### Filtering

Like a gold prospector panning for gold, we need to pan for happiness, solutions, enlightenment, wisdom learning, fulfilment and progress in the rush of information that flows through our life.

### Intersecting

Build upon solid bedrocks of experience. The most valuable thing you can have is congruence between your intent, your actions and your thinking. Look for the things that are consistent, despite changing context. Seek the solid stepping stones where they intersect. Recognise the paradox of growth.[56]

### Connecting

Focus on the invisible web that links ideas. Practice unconstrained thinking. Put yourself at the point of creativity. Refuse to choose the lesser of two evils, instead harvest the strength of merit wherever you find it. Fill the blanks with positive behaviour.

---

56 See the next section "The Paradox of Growth"

# Evolution Of Knowledge I

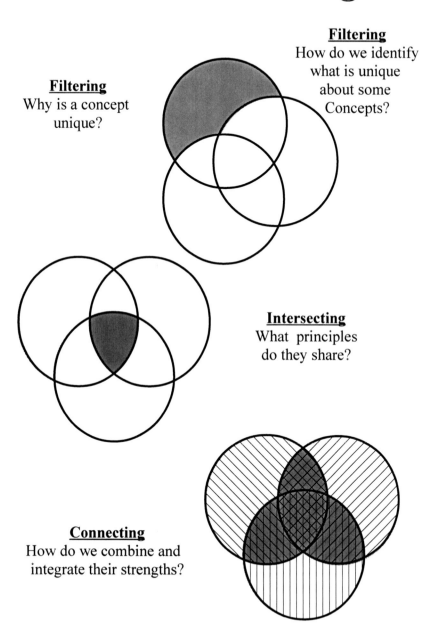

**Filtering**
Why is a concept
unique?

**Filtering**
How do we identify
what is unique
about some
Concepts?

**Intersecting**
What principles
do they share?

**Connecting**
How do we combine and
integrate their strengths?

# Evolution Of Knowledge II

**Learning**
Isolating something
and filtering it so
we can see what
it definitely is.

**Understanding**
Examining what it
has in common with
all the things we
think we know

**Applying**
Testing the concepts
across every possible
combination that we can imagine
and observing why they work,
what they change and how far
and where they
take us

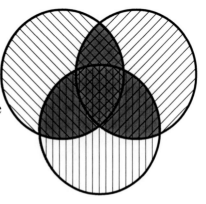

# Evolution Of Knowledge III

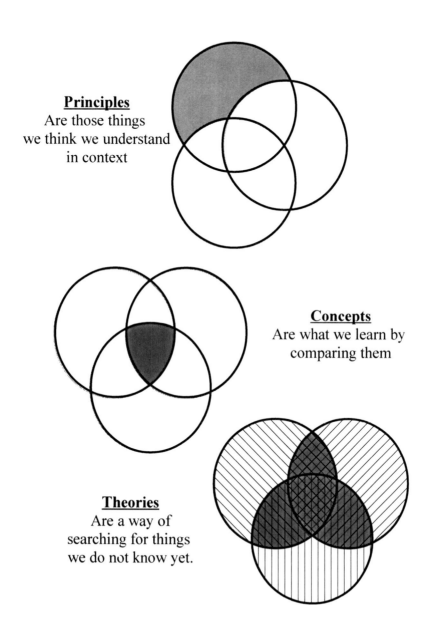

**Principles**
Are those things we think we understand in context

**Concepts**
Are what we learn by comparing them

**Theories**
Are a way of searching for things we do not know yet.

# The Root Of Knowledge

> *"Science always has its origin in the adaptation of thought to some definite field of experience."*
> **Ernst Mach**

## Thoughts

There is a paradox at the heart of reality and at the heart of every thing[57] that exists. This includes ideas. It not only includes ideas, but it starts with ideas.

Ideas exist. I think therefore I exist, or, at least, thoughts exist. Something seems to exist anyway. Maybe existence is just the echo of thoughts; maybe thoughts are just the echoes of existence. What do I know? What does anybody really know?

Well, I know for certain that if I start talking like this to many business clients, I might as well collect my coat and leave.

As far as we are concerned, our current reality is a system defined by time and space. As far as we can determine, because we are part of the system, time and space are its boundaries and therefore the boundaries of reality.

The fact that we do not talk about it at the office, does not change the fact that everything hinges on a thought. This is still true, whether that thought instigates the action or

---

[57] "Thing" is my favourite word. "Thing" is the unanchored metaphor. "Thing" says that there is this discrete pocket of existence but you remain agnostic about its qualities and you do not judge it. "Thing" says that something exists and that is all you have to say about it. "Thing" says that you choose not to create presuppositions. "Thing" is the uncarved block until you choose to carve it carefully into something and give it shape in your universe of perception.

whether that thought is the rationalisation of a deterministic universe.

The thought is all we can really be sure of. It is our experience of existence. Even if we are all in a large computer simulation somewhere[58], even if we are all characters in the Red King's dream or even if the world of matter is all there is, thoughts are our interface to reality, our sole point of contact.

All of your other senses communicate with your consciousness through thought.

Thoughts are electrical impulses in the brain. They are organised energy. Energy is matter at the speed of light squared. Matter exists in time and space and determines our course.

## The Paradox Of Thought

Thoughts are also paradoxes by their very nature. The thought to raise your finger, say a word, caress a loved one or

---

[58] Hilary Putnam in his 1981 book (Reason Truth and History) updated Descartes' Evil Genie from his 1641 "Meditations on first philosophy". Descartes asked how would we know if the evidence of our senses was not just the illusion of some malevolent godlike creature fooling us. This led him to his famous "Je pense, donc je suis". Putnam proposed brains in vats being fed sensory information. This idea was visualised in the film "The Matrix" in 1999. The philosopher Nick Bostrom suggests that it is likely that in the future we will be able to create hugely realistic virtual worlds and virtual minds. We may all retreat to them since the resources required to maintain them will be relatively tiny. He goes further and hypothesises that there will be vastly more simulated minds and that they will all consider that they inhibit reality. He suggests that this future may already have happened and that the numerical odds are in favour of us already being in such a simulation.

take a step appears to take place a fraction of a second[59] after the command impulse has been sent to the muscles.

Unsurprisingly, this has frightened a lot of people. It is a bald fact and, without context, it seems to say something very odd about free will.

Dig a little deeper and it is not quite so bald. Dig a little deeper still and it seems to say something even more strange about consciousness. It seems to inform us that consciousness is recursive. Let me expand on that.

In the realm of the very small – the realm of quanta – time, cause and effect are all rather more arbitrary than is comfortable if we want the world to be predictable and conservative. Ironically, it appears that it may be all about prediction and conservation.

Our brain predicts the future – we see the world as it will be a fraction of a second in the future. It goes to a great deal of trouble to edit out anything that does not conform to that prediction. There are a range of behavioural experiments that indicate this. Among them is one that demonstrates that we will not see a gorilla in the midst of a basketball match[60] if we are not looking for it. Another demonstrates that if the person interviewing us changes mid interview, our brain will not bother us with the information[61]. Our brain does a lot of work to provide us with a seamless reality.

---

59 Bereitschaftspotential or pre-motor potential or readiness potential was first recorded and reported in 1964 by Hans Helmut Kornhuber and Lüder Deecke at the University of Freiburg. They showed, and it was later verified by a series of experiments in the 1980's by Benjamin Libet, that there is activity in the motor cortex from .3 of a second up to 1 second before voluntary movement. Simply stated, this means that there is activity in the brain leading up to movement before you consciously decide you are going to move. This phenomenon is now being used in brain-machine interfaces in systems that allow thoughts to direct computers and prosthetic limbs. Watch this amazing video by Krishna Shenoy, Stanford University. http://www.youtube.com/watch?v=I7lmJe_EXEU

60 Daniel Simmons and Christopher Chabris have written "The Invisible gorilla" about their research into this phenomenon. www.theinvisiblegorilla.com

## Reality Is A Magician

Imagine watching a magician performing. The magician does a piece of magic. You don't know how it is done. You know it is an illusion because there is no such thing as magic, is there? This knowledge, that it is all an illusion, allows you to suspend disbelief and helps you to go along with it, in order to be astonished and amazed.

So, the knowledge that there is no magic allows you to temporarily believe that there is magic and enjoy the trick.

What if there were someone who really could do magic? What if they realised, that in order to perform, and for you to enjoy their show, they must set up a trick that also includes a rational explanation?

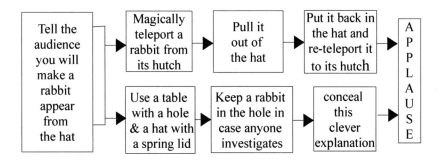

You are satisfied that no rabbits were created out of thin air, although that's the way it looked. Because we know that there is a rational explanation available, the magician got to do magic without being burned at the stake or dissected in a laboratory.

Well it appears that reality is quite the magician.

---

61 This is known as "change blindness". There is a paper on it here http://citeseerx.ist.psu.edu/viewdoc/download?doi=10.1.1.7.9990&rep=rep1&type=pdf. :"Evidence for Preserved Representations in Change Blindness" by Daniel J. Simons,1 Christopher F. Chabris, and Tatiana Schnur of Harvard university and Daniel T. Levin of Kent State University

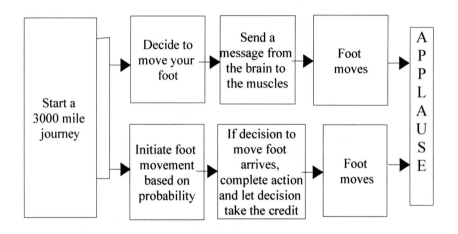

## The Rational Explanation

One explanation I really rather like is that evolution has equipped us with a predictive ability and a sensory delay to allow things to appear to happen in real time.

Think of it like this: if pre motor potential did not exist, then that delay of up to a second would appear to happen between the decision to move the foot and the moving of the foot. Result: the brain would be constantly lagging behind reality and unable to deal with real time actions.

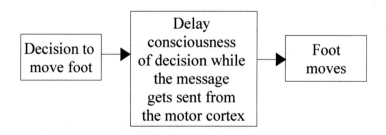

How I see it is that for evolutionary reasons we need to feel that our decisions and actions are happening in real time and therefore the decision and the action are arranged to be parallel in time and instantaneous.

It gets even cleverer. There is a suggestion that the delay is there in the first place, so that, when you decide to move your foot, the conscious decision is delayed until there is feedback that the foot is, in fact, moving. The model is delayed to coincide with the reality.

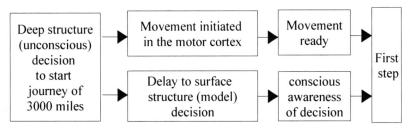

This still does not answer the question of whether there is free will or not. It appears to be illusory because events start earlier than decisions. Why then have we been left by evolution with a subjective sense of free will and a delay in decision to move our foot until it is actually moving?

I propose that it is the magician of consciousness providing a rational explanation because we do not really want to look into a fabric of reality constructed of probabilities and unconscious motivation.

Our conscious mind needs to believe that it is in control – consciousness depends on the illusion of control (rational explanation for magic trick) which probably masks the reality of control (the deep magic that makes it happen). It is entirely possible we have a deep structure, unconscious-consciousness, in touch with reality, and a surface structure, rational-consciousness, driving the model of our choice. So how do we choose the model?

Let's look at the three simple ideas proposed on page 120[62].

---

62 Filtering. Intersecting. Connecting.

# Filtering

*"Life is not a problem to be solved, but a reality to be experienced."*
Soren Kierkegaard

## Context

The difference between ideas and thoughts is that we tend to categorise thoughts into ideas. An idea has a consistency or purpose about it. It has an outcome and consequence associated with it, even if it has been spontaneous. An idea tends to be a peg to hang thoughts on. Ideas become models.

## Filters

What defines anything as itself? How do you take what is useful and what do you do with the rest? How do you separate the baby from the bathwater? How do you know if the bucket is spinning around the water or if the water is spinning within the bucket?

## Mach's Conjecture

One of the strange thought experiments in physics is Mach's conjecture[63]: "mass out there influences inertia here."

The question was posed in a metaphor using stars and centrifugal force. It basically asks how you know you are moving and what forces are really at play.

> Have you sat in your car looking at the car in front and let your mind drift while stopped at traffic lights? When the car in front starts to move off have you ever for a second felt that

---

63 This is generally attributed to Ernst Mach 1838-1916 Philosopher and physicist who gave his name to the speed of sound.
http://plato.stanford.edu/entries/ernst-mach/
http://en.wikipedia.org/wiki/Ernst_Mach

> you had started to move backward making you reach for the brakes? In this situation, because we momentarily lose reference, we are not sure which car is moving.

Let me put it another way. You are only moving because of your relative motion to something else. Now imagine you are alone in the universe. There is nothing else, just you. Are you whizzing along at a fantastic speed or are you staying absolutely still? Is there any way to know and in an infinite universe, is there any difference at all? Would even inertia[64] mean anything? The only way we are aware of motion, or non motion, is through our detection of gravitational forces. Without any other gravitational masses would centrifugal force exist even if you started to spin?

It is only relative to other objects in the solar system that we seem to be moving. We are orbiting relative to the sun. The sun is travelling along relative to the galaxy and the galaxy is hurtling along relative to the other galaxies in the universe and those galaxies themselves appear to be flying apart.

This concept was one that interested Einstein and affected his theory of relativity. He said that Mach could be considered the precursor to the general theory of relativity.

Only within the framework of the rest of existing matter can we know what is rotating.

The frame of reference is everything. It is our context.

## Boundaries As Context

We only know who we are by measuring ourselves against what is around us. We are constantly filtering differences. We are constantly determining boundaries. Boundaries define the essence of a thing. Anything is only what it is, within a framework. Context relies on identifying where we are in relation to something else. It relies on our ability to filter.

---

64 Resistance to state changes – stillness to motion, slower to faster, faster to slower, one direction to another.

We filter, we categorise and we file. We are difference engines.

## Atomic Distinctions

Knowledge is about the decisions of where boundaries exist in information. It is about finer and finer distinctions. The more you know about something the finer the distinctions you can make. A master musician is one who can detect the finest of distinctions in sounds. A master architect is one who can detect the finest distinctions in shape, form and material. A master chef can make the finest distinctions in flavour, texture and taste. Gaining knowledge is about refining distinctions.

Facts may appear to be atomic, but facts are themselves refinements from a sea of observation. The more refined and pure they are, the more useful they are. They are also only defined, as motion is, within a frame of reference. Facts become blurred with time and context. They change.

## Knowledge Is A Product Of Evolution

Mach's other speciality was the philosophy of science. He believed that knowledge is a product of evolution. Simple experience becomes *a-priori* knowledge which enables further more complex experience. Complex understandings can be built on this process.

He said that "A piece of knowledge is never false or true – but only more or less biologically and evolutionary useful. All dogmatic creeds are approximations: these approximations form a humus from which better approximations grow."

Nothing exists in isolation. Knowledge is interconnected with other knowledge. It only makes sense in a context of facts and may change its meaning in another context of different facts.

## Filtering And Curious Water

We live in the information age which sometimes feels more like the information overload age. We are swamped with information everywhere we turn. We used to think that we did not have enough information. Now we have more than we could ever want.

We swim in a river of information which runs through our gills like water. We take meaning from this information as a fish takes oxygen from the water. If the river is fast flowing, the fish can appear stationary as the current carries water through its gills. If the current has stopped, the fish must swim to create the flow. It is all a matter of perspective. To appear stationary in flow the fish must swim ferociously against the current. The fish that appears to you to be moving in a still pool may not be putting as much effort into swimming as the one holding stationary against the current that is invisible to you.

It is the same for knowledge workers. Sometimes to maintain position against a torrent of information demands extraordinary skill and manoeuvrability. Sometimes to keep the organisation alive they need to generate movement in a stagnant environment[65].

What is this information? How do we know what is oxygen and what is industrial waste?

When we think of filtering we imagine it as a process for removing impurities or a process for isolating values that fit through the mesh. In some kinds of filtering, something else entirely happens.

---

65 This is a tricky one to call because the shameless self promoter will often make sure that they stay out of the rapids. They point to the apparent lack of progress of those holding position against the barrage of change. They point to their own excellent productivity within a stagnant process while they sun themselves like big fish in a small pond.

*As part of her science education, my daughter and I began to investigate the de-ionisation[66] process for removing unwanted metallic ions[67] from water. As we studied it, we discovered something unexpectedly surprising and interesting.*

**Water to be purified is run through an exchange membrane. The metallic ions react chemically and attach themselves to the membrane material, forcing it to releases hydrogen ions.**

**E.g. Sodium ions ($Na^+$) combine with the membrane and displace one ion of Hydrogen ($H^+$). Calcium ($Ca^{++}$) displaces two and Ferrous ($Fe^{++}$) displaces three hydrogen ions into the water. In other words, the poisonous substances "stick" to the resin and dislodge hydrogen ions to take their place in the water.**

*This is a wonderful example of practical chemistry that we both enjoyed reading about. It dislodged lots more questions about ions, atomic structure, pH values and the properties of water. These questions, themselves displaced ions of knowledge that hung about us and searched to stabilise and balance themselves with answers while we pressed on with de-ionisation.*

**At this first stage of de-ionisation the water could be considered dangerous. Because of the $H^+$ ions, it can now donate in solution, it has become acidic. It is unbalanced. If we were to stop here, the process would have failed on the basis of the superficial evidence. We started with balanced water which contained dissolved contaminants. Now we have unstable, acidic water.**

*A little bit of knowledge is a dangerous thing. It can become unbalanced. That is why this was a good time to stand back*

---

66 The Usborne Science Encyclopaedia page 73 and
http://www.freedrinkingwater.com/water-education2/49-water-di-process.htm

67 An ion is a molecule where the number of electrons is not equal to the number of protons thus creating a positive or negative charge. More electrons than protons and you have a negative ion, less electrons than protons and you have a positive ion.

*and tackle some of those questions we dislodged about the basics of chemistry and water.*

*Water is slightly curious. It is the most important substance on the planet. It is the medium for the existence of life. We are two thirds water ourselves. Chemically it is unique. It is the only substance on earth that can exist in all three states of matter (solid, liquid and gas) in the same place, time, pressure and temperature. It is the only substance that expands as it gets colder. It has the highest surface tension of any common liquid with the exception of mercury*

*Its molecule ($H_2O$) has two Hydrogen and one Oxygen atoms, bound together by sharing pairs of electrons. This is called a covalent bond – the electrons travel about both nuclei.*

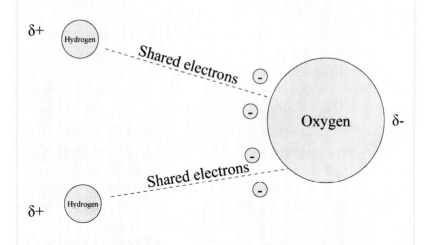

*The very curious thing about water is that it is a polarised molecule. One side is a little negative and the other side is a little positive.*

*The Oxygen atom is more attractive to the electrons than the Hydrogen in each bond. This means that the oxygen end has a little more negative charge and the Hydrogen end has a little more positive charge. This polarisation gives water some of its most interesting properties.*

> *You could say that it knows it is a little off balance and is trying to elicit feedback from its environment. It is a kind of chemical curiosity that makes the molecules of water crowd around any ionic or polar compound (e.g. salt, ink, carbon dioxide) that enters it.*
>
> *This curiosity may sometimes not be convenient to us as water can get very interested in the above mentioned undesirables such as Sodium (Na), Calcium (Ca) and Iron (Fe). So let's get back to the de-ionisation.*
>
> **There are too many Hydrogen ions floating about so the acidic water is passed through another resin membrane. This membrane releases negative hydroxyl ions or anions[68]. These anions combine with the hydrogen cations to create water molecules. These are indistinguishable from the water in which they had been made.**
>
> *The questions dislodged by our investigation were answered for now. We had a new understanding of water and its properties. We were starting to understand ionic and covalent bonds and we were gaining some deep respect for water's chemistry. The information was integrating with what we already knew. The ions were fast becoming knowledge.*

When we have unnecessary detail poisoning our clarity, what if we could use an information membrane? What if we could have an information de-ionisation process that would clarify what else we need to know and identify what internal prejudices were preventing us from learning? What if we could have a filter that allowed us to remove the threat and see the potential in new information?

## The Two Faces Of Filtering

Filtering is not just removing unrelated knowledge so that we can get a clear view of the piece we are interested in, it is also recognising what needs to be added to a piece of knowledge to stabilise it and make sense of it.

---

[68] Negative ions are called anions. Positive ions are called cations.

There is an effective form of study in which you draw a map of what you know and then identify the gaps that represent the things you need to know in order to understand. The idea of understanding is central to filtering.

Filtering is the first stage in understanding. Understanding is an iterative process that requires intellectual honesty. You filter, not to get rid of uncomfortable information, but to highlight what you might be missing and to dismiss false connections, complex equivalences and limiting beliefs. With these out of the way, the path to enlightenment is often clearly signposted with question marks.

# Questions Are Filters

> *"What makes the desert beautiful is that somewhere it hides a well."*
> Antoine de Saint-Exupery

## Framing The Problem

Lots of aphorisms are attributed to Albert Einstein. Apparently[69] he was rather bemused by this and commented "In the past, it never occurred to me that every casual remark of mine would be snatched up and recorded. Otherwise I would have crept further into my shell."

From what is attributed to him, despite his modesty, he was clearly a master of imagination, intuition, humanity and of questions. There are several accounts of his explanation that if he was given an hour to solve a problem upon which his life depended, he would spend 55 minutes of that hour searching for the right question to ask in order to frame the problem properly.

## Process Grows

If we take the metaphor of executive function based on decisions and legislative function based on the precedent of repeated behaviour, we can start to see a pattern.

This applies to individuals whose unconscious is the legislative function and whose conscious is the executive function. It also applies to countries, to organisations and to computer programs.

In many organisations, these rules become **The Process**. Process can quickly become a giant hairball of rules. Each rule makes sense in isolation, but as more and more of them are applied in the same place, the process becomes a tangled

---

69 www.guardian.co.uk/books/2005/apr/02/featuresreviews.guardianreview36

mess. This mass has gravity and inertia. There is some sense in learning how to orbit it[70].

## Programming Precedent

Computer programs are interesting to analyse because they do what they are told. A programmer begins to code based on customer requirements and the features of a programming language. These things create precedent.

The programmer is the executive function and the program is the legislative function. If the requirements become confused, the program turns into a sort of hairball of conflicting laws.

If progress is made against misunderstood requirements and subsequently fixed on a superficial level, the code remembers. Just like an unconscious mind, it holds impressions of what was intended when it was written, deep in its logic.

Anything which springs from consciousness behaves like consciousness. This is because it has been an expression of consciousness and a mirror of the mind that created it.

The program behaves not unlike a mind with a short term conscious and a long term unconscious memory.

## The Programmed Mind

The conscious mind forges ahead while the unconscious observes, records, advises and squares perception and reality.

This is a pretty good scheme. We acquire knowledge and bank it. We compress information into patterns. We can call on our experience and learning to get us out of tight spots without having to think it out every time.

We think about what we want to say and the words just arrive with all their compressed meanings. Instinctive reactions in sport, profession, self preservation and communication, all

---

[70] For instructions see "Orbiting the Giant Hairball" by Gordon Mackenzie

rely on the legislative function to provide us with strategies to function effectively. Sometimes we can consider our action and sometimes the legislative function is so sure of the mind of the president that the rule is applied automatically without recourse to the executive function. As we swerve to avoid the child chasing the ball out onto the road who cares what the president thinks?

## Living In Free-Fall

In order to allow us to function in the face of the immensity of a reality that we do not understand and cannot comprehend, the legislature of the unconscious has become adept at distortion, deletion and generalisation.

### *Distortion*

There are gaps in our understanding. So that we can get on with living, we distort reality and it appears seamless to us. Otherwise we would be stuck in its twin headlights of eternity and mortality like stricken rabbits.

### *Deletion*

If we had to stop to examine every anomaly or interesting diversion in our path, like a two year old who delays the trip to the circus to examine the cracks and weeds on the front doorstep, we would be run down by the locomotive of time before we got a good look around. Therefore we delete detail so that we can see a bigger picture and move about.

### *Generalisation*

So that we can move on and discover the new, we are brilliant at pattern recognition and generalisation. Volume one examined this as classification and abstraction.

### *Extrapolation*

Communication is a matter of extrapolation. Just listen to any two people talking and note how many sentences are unfinished and are accompanied by shrugs, "you know" and "like". We expect other people to get on our wavelength and extrapolate.

### *Approximation*

It is strange but true that these flaws in our communication are the reason it functions. We do not describe everything. We do not specify everything. We do not check and verify.

We use approximations, guesses and intuitions all the time. We give incomplete instructions and descriptions. It is in these gaps that the real communication takes place. We fill the gaps with something akin to mind-reading. We do it so instinctively and frequently that we consider it pedestrian.

## The Human Advantage

Our legislative body is a dreamer. It has to be. Where there are gaps in knowledge that we cannot wait to acquire, it fills them with beliefs, assumptions and hallucinations.

This is abundantly clear in the search for Artificial Intelligence. Human consciousness ignores the gaps in reality and spans them with beliefs it makes up. Machines have no imagination. Machines do not believe. Machines cannot believe.

The concluding chapter of volume one asked *"What if reality were an infrastructure upon which we lay our theories and our dreams?"* It proposed that *"Those principles which are laid upon the solid infrastructure hold fast. Principles laid upon those adhere. Principles laid across the spaces obscure the light."*

Consciousness is what spans those gaps in the superstructure of reality.

> There is a joke about two inmates escaping an asylum across the rooftops at night with a torch. They come to a drop between buildings and almost despair. One then has a bright idea and shines the torch across the gap. He invites the other one to walk across on it. "Oh no", says his friend, "you might switch it off when I am only half way across."

Similarly there are times it is best not to examine reality too carefully until you are safely across. We can call these beams of light that we walk on, axioms or presuppositions. They do not need to be true in every absolute sense[71]. They just need to be internally consistent with our current reality. We need to be sure that we are not suspended on them in mid-air before switching off their generator.

## The Two Faces Of Versatility

It is my axiom that what is very useful also has to be very versatile[72]. In being flexible enough to be useful something will have the potential to be counter productive and dangerous. It is a matter of counterbalances.

This applies to the versatility of consciousness.

Mistakes and misunderstandings can become rules. We can delete, distort and generalise, extrapolate and approximate vital information.

This, then, is where we came in. In all human communication there are elements of distortion, deletion and generalisation. Students of linguistics have grappled with this fact since we first started to study language, communication and thought.

---

71 Indeed, as Einstein observed, there is no absolute time or space, which tends to make any other absolutes a bit tricky.
72 See "response ability" on page 193

## Working Models Are Our Speciality

My other axiom is that methodologies work very well. Any of them can give you what you ask for. For a project manager, that is not the issue. Most problems arise because we are convinced, one way or another, to do things that we really do not want to do. We agree to do what we think we should be seen to do and hope to massage it into what we really want.

As time goes by we become more and more afraid to admit that we did not really ask for what we wanted. We conceal from ourselves that we did not communicate what we wanted to do. It is easier to blame the technique, the methodology or the process.

When we have decided that we are not interested in the details, we construct a big picture. This is a recipe for deletion, distortion and generalisation. We are very talented at making big pictures out of very few details. This is fine as long as the details we chose to extrapolate were ones we understood and had correctly interpreted. Our versatility is that we can create amazingly convincing models out of supposition and excuses as well as out of observation and evidence.

## Asking For What You Want

Computers do not interpret[73], extrapolate or guess in the way human beings do. That makes them merciless judges of what you said rather than what you meant. They give you exactly what you ask for and there is no getting away from it. It is black and white.

So asking for what you really want is the thing to do, but people rarely do. They have a variety of reasons for this strange behaviour. They may not know why themselves. They

---

73 Semantically incorrect but I am talking about a different type of interpretation – lateral interpretation. I am aware of Perl, Python and Ruby as program interpreters. Strictly speaking they are translators that use fixed rules.

may be trying to take part in the litigation lottery[74]. They may be trying to keep their options open. They may be asking for what they think you want to give them. They may be second guessing your second guess. They may be asking for one thing and hoping you will give them what they want by mistake and as it's not what they asked for, they won't have to pay. *The other game* gets convoluted, believe me.

## Filtering Tools

NLP is based on the art of asking questions that uncover, clarify and disambiguate. It has tools to get from what you think you want to what you really want. When you make statements, it encourages you to examine how you know what you think you know and if you really know at all. At its best it encourages you to achieve a perspective where you identify consequences.

In terms of project methodologies, traditional waterfall is great at giving you exactly what you ask for. It is not so good at making you think about whether you need it or not and it always lies about time and effort. Agile, particularly Extreme Programming, is based on asking the right questions. Test first programming is all about crafting elegant questions. A good test is a question that makes the answer easy. Einstein's 55:5 ratio applies.

To make this work you must make use of embedded questions that reveal the intrinsic value. These embedded questions are integral to certain ways of working and dealing with each other. They are the product of the natural curiosity of people interested in what they are doing. Some examples might be:

1. The questions between programmers working together.
2. The test as question in test first/test driven programming.
3. The questions between stakeholder and technical teams.

---

[74] Slipping in sneaky requirements and clauses that will cause the supplier to default and pay out penalties of some sort – maybe cash, maybe guilt.

Technical challenges are only a matter of time and expertise. Communications challenges are a matter of formulating useful questions.

# Intersecting

*"There are three classes of men –
lovers of wisdom, lovers of honour, lovers of gain."*
**Plato**

## Balance

Ernst Mach also discovered how the sense of balance works. It manages the feedback information received from liquid in the inner ear.

The system of balance is known as the vestibular system and can probably be added to the 5 senses of sight, hearing, touch, smell and taste, listed by Aristotle. In a very real way the sense of balance is the deep structure of our other senses.

The sense of where we are in space and the perception of space is the parent of all other senses. Without it, hearing, sight and touch might be just so much confusing input with no meaning.

Balance works by intersection. Liquid moves through three semi circular canals in the inner ear and the triangulation of feedback from these gives us a sense of where we are in three dimensional space.

## Depth Of Perception

I listened with interest as Sue Barry[75] described discovering three dimensional sight. Her discomfort at discovering sharp edges and corners; her wonder at the folds in cloth and the spaces in the world, is still palpable almost a decade after acquiring three dimensional sight.

More interestingly she described people who have always had stereoscopic vision telling her that it is possible to have 3D

---

75 http://www.scientificamerican.com/article.cfm?id=seeing-in-3-d

vision without two eyes. Their reasoning is that if they shut one eye they still see in 3D.

I smiled as she, a neuroscientist herself, explained that this is because their brain has a lifetime of 3D images stored. This allows it to deduce what this image would be if it were to be three dimensional.

## Triangulation

Both hearing and vision rely on the intersection and triangulation of input. This provides information as to where things are in space and thereby work with the vestibular system to provide us with a sense of existing in space.

You need three points to triangulate. In the third century BC the distance of the moon from the earth was calculated by the astronomer, Aristarchus, using triangulation.

Intersection relies on three point triangulation.

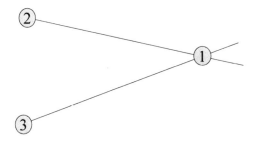

I have noticed the evidence building around a ubiquitous rule of three and how we rely so heavily on the intersection of threes.

## Triangulations And The Power Of 3

### *Some famous threes:*

- Three Dimensional Space
- Id Ego Superego
- Mind Body Spirit
- Love Wisdom Order
- Model View Controller
- Axis: X Y Z
- The 3 Musketeers: Athos Porthos Aramis
- Height Width Depth
- Proton Neutron Electron
- Three Laws Of Motion
- Veni Vidi Vici
- Wine Women Song
- Three Laws Of Thermodynamics
- The 3 Stooges: Curly Larry Mo
- Three Laws Of Robotics
- Faith Hope Charity
- Red Amber Green
- Thesis Antithesis Synthesis
- Understanding Judgement Reason
- The Good The Bad The Ugly
- Liberté Égalité Fraternité
- Yes No Probably
- Authority Achievement Affiliation

Three seems to be an important quantity from Aquinas to Hegel and Kant, from the Bible to Asimov's robots and from the rules of rhetoric to quantum theory. To me it is more obvious than mystical.

Although we are bipeds we know that stability comes from three points. We know that navigation relies on three points. We know that the lever depends on three points, the force, the load and the fulcrum.

Achievement, affiliation and authority were proposed as the basic human motivation by David McClelland based on the work of Henry Murray who in turn was influenced by Alfred North Whitehead.

These three A's cover a lot of what we do and why we do it. We want to get things done for their own sake, we are social creatures and we want to determine our own destiny.

When it comes to almost anything, we like to have three of them. In NLP we often observe that many people need to see, hear or feel things three times before they are convinced.

### *Why three exactly?*

The simple answer is that we need a third point for perspective.

In working and learning groups, three is a very useful number: two people to carry out a transaction and a third person to act as an observer. This third person is often the coach.

With three we can start to deduce interesting information from intersections. Intersection is the point of balance.

# Connecting

*"Principles for the Development of a Complete Mind:
Study the science of art. Study the art of science.
Develop your senses – especially learn how to see.
Realize that everything connects to everything else."*
**Leonardo Da Vinci**

## Inspiration

The moment of connection is the culmination of all that has gone before. We load our mind with experience as we travel through life. The richer and more varied that experience is, the more connections we are able to make.

Although it often seems that inspiration comes crashing unexpectedly out of the void, it surfaces breathless out of our unconscious mind like a pearl diver dragging a net of connected ideas from beneath the waves of thought. The contents of the net depend on how well we have seeded the beds of our imagination with information and how deep and long we have trained the diver to go.

I think this is what Edison meant when he said that genius is 99% perspiration and 1% inspiration. That one percent moment makes the ninety nine percent of other moments worthwhile. It is the coalescent thought that unites experience, theory and action.

## Peak Experience

It is in that one[76] percent that we have the advantage over computers, process and artificial intelligence. It is not an algorithmic process or even a heuristic one. Inspiration is irrational and illogical. It gathers disparate facts, metaphors,

---

76  It is clear to anyone who cares to read about great inventors, scientists, musicians, artists or writers that it is a great deal more than a one off 1% inspiration. Inspiration is like lightning and will strike repeatedly where it has found an easy passage.

dreams, visions, desires and beliefs then connects them. It merges and connects the seen, the heard and the felt, all in a glorious synaesthesia. The brain seems to have a peak experience[77] during which it connects the conscious and the unconscious seamlessly.

It may be that because the brain holds all information in the same format that it can do this. Moments of pure inspiration feel more like a sigh and a momentary relaxation of vision. Something just moves into the frame, so clear that you are amazed you never noticed it before.

## Connection Is The Mechanics Of Inspiration

Connection depends on the quality or purity of the content (i.e. the filtering) and the integrity and honesty of the intersection. This is what has entered the unconscious as important and valuable. This is the fabric of which your model is constructed.

We can return to our circles and planes where we discovered that each circle can rotate in an infinity of planes through any point into an infinity of infinities of infinities. Using this metaphor, inspiration is when some part of our consciousness imagines the sphere that would result if the circle were to be rotated on every possible plane through any diameter. It

---

[77] Abraham Maslow "The Further Reaches of Human Nature".

begins to admit the possibility of the intersections in other planes.

If the original circle is your model of the world containing the problem, the usefulness of the inspiration that is the resulting sphere, that encloses the planes, depends on how much you have stretched your model to accommodate the evidence of your senses.

Inspiration often ignores the limitations of one model and connects to what is possible in other models. If the metaphor I have been using in this book were to be drawn, it might look something like this as the different models collapse into a perception that accommodates them.

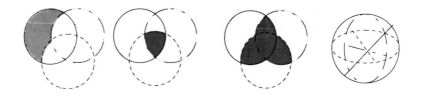

Models which can seem to be diametrically opposed when viewed from their own planes, collapse into a sphere at moments of inspiration and everything seems to be connected at once. The tenuous connections take hold and alter the reality of the observer.

This is rather like visualising two dimensional representations rotating into three dimensional space. Connections that seemed impossible between models in the two dimensional become inevitable and natural with three dimensions to work with.

It is worth remembering that the mind has more than three dimensions to work with. The mind can make connections across time, space, discipline, emotion and any temporary boundaries we wrap subjects up in.

Leonardo Da Vinci, when confronted with problems that challenged even his formidable creativity and intellect, knew this. When he was faced with roadblocks in one sphere, he looked at solutions to something else in another and allowed his mind to connect to an answer for the original problem. He knew that there is no such thing as an intractable problem, just an unhelpful perspective.

We often need to distract ourselves from the boundaries we have set and the strategies we have been practising. Quite often we think we are exploring all the possibilities when we are simply pacing the confines of our cell.

> *"As he paces in cramped circles, over and over,*
> *the movement of his powerful soft strides*
> *is like a ritual dance around a center*
> *in which a mighty will stands paralyzed."*
>
> *Rayner Maria Rilke (1875–1926)*

Inspiration involves making connections in order to shatter the bars of limiting beliefs, free the mighty will and remake the world.

Connections are everywhere. Leonardo was an artist, a sculptor, a scientist, an architect, a linguist, a botanist, an anatomist, a musician, a mathematician and an architect. When you start to study him you see that he could move the boundaries at will, that he embraced paradox and that he presupposed connections.

## Breaking Eggs To Make Omelettes

### *Filter*

Isolate what is unique and important in each model.

### *Intersect*

Identify where all models agree or where they would agree if they did agree.

### *Connect*

Allow connections which are outside your model of the world.

### *Build a better model*

The resulting inspiration may shatter your model but it will rebuild a better one.

## How To Connect

Read books, listen to talks, explore art, engage with people and ingest newspapers and magazines that all challenge and expand your opinions and models of the world.

Engage with differing viewpoints as more than a scouting exercise for a battle of ideologies. Treat it as the exploration of an exciting new world, among whose dangers great discoveries await to be made.

Seek out new ideas that will shatter your model of the world.

When the old model shatters, you will retain all of its experience and value. When you are forced to filter old knowledge in light of the new information, look for points of intersection and connection.

Integrate what you know to be true and valuable from the old with your new learning. The new model becomes your new old-model. Everything you loved from the old model will be enhanced in the new model. Trust me on this.

Remember, when in free-fall through a shattered model, look for evidence. Sometimes you will be presented with choices that you will need to investigate quite carefully in the light of your new understanding. They may have previously been mutually exclusive, *either-or*, but new knowledge may present you with a new perspective that changes the boundaries and allows you to integrate them.

It is useful to keep this in mind when presented with *either-or* choices in complex situations. Many *either-or* choices are limiting, especially if you suspect that you are being presented with artificial constraints and conditions.

> It is like the old chestnut: "Have you stopped beating your wife yet? Either you have or you have not. Yes or no." Reframing the question as being about your prowess at chess or being able to assert that you never beat her in the first place gets rid of the *either-or* and the implied presuppositions.

Even in science *either-or* encourages you to belong to a particular school of thought and to highlight only differences and ignore connections. Subsequent discoveries tend to highlight that a whole stream of breakthroughs were just below the surface ready to shatter both models and rearrange the facts. (E.g. light as a wave or a particle, critical periods or brain plasticity etc.)

When you find that you are under pressure to make black and white choices it is usually an indication that both models are incomplete, particularly if you have to ignore compelling evidence either way you choose.

If something is defined by what it is not, it is in denial. It is only offering you a limited choice of models and perspectives. An *either-or* choice like this should be a trigger to look for connections and they are usually there to be discovered

Embrace the possibilities and welcome the discoveries that will allow you to connect to the outer orbits of wisdom.

Seeking to connect knowledge creates a net for inspiration, understanding and innovation.

Inspiration is a function of searching with an open mind for connections across models and being one of the people who can see, or who can see when they are shown[78].

- ✔ Filter
    - ☑ Be curious: ask questions and create clarity
- ✔ Intersect.
    - ☑ Seek experience: test your understanding.
- ✔ Connect.
    - ☑ Seek uncertainty: welcome paradox as the invitation to learning.
- ✔ Context
    - ☑ Refine your senses: observe relativity.
- ✔ Balance
    - ☑ Integrate your skills: connect the areas of your life.
- ✔ Inspiration
    - ☑ Recognise interconnection: create new patterns.

Connecting is recursive. It contains filtering, intersecting and itself, just as inspiration contains context and balance.

## Einstein's Brain

Einstein had the same neurobiology as you. So did Leonardo. When they dissected Einstein's brain[79], they found only three things of note. It was a bit smaller than average, size is not associated with intelligence. It was missing the parietal

---

[78] Leonardo said: "There are three classes of people: Those who see. Those who see when they are shown. Those who do not see."

[79] "The Exceptional brain of Albert Einstein" Sandra F Witelson, Debra L Kigar, Thomas Harvey published in the Lancet 19 June 1999

operculum, a small wrinkle. The inferior parietal lobes, associated with visual and mathematical thinking, were a bit enlarged.

Through what we know about **"use it or lose it"** from neuroscience, and what we know about Einstein's habit of thinking, it is not surprising that Einstein might have enlarged these areas. Who knows what he started out with? His parents thought he was retarded when he was a small child because he did not talk[80].

Although these remarkable people from history may seem to us to have been unusually gifted, they were operating with the same tools as ourselves.

## The Prepared Mind

It is interesting to read about great breakthroughs in science. They mostly start with a phrase like "That's odd!" followed by "I wonder. What if?"

From Archimedes in his bath, Marie Curie and the fogged photographic plates, Galvani with his frog legs and Fleming and his mould[81], "In the fields of observation, chance favours only the prepared mind."[82] While all of them had moments of

---

80 Apparently he seldom spoke and when he did it was in complete sentences which he had practised under his breath. This went on until he was nine. Thomas Sowell has explored a tendency of brilliant people to develop speech later in childhood. It was noted in others such as Feynman and Arthur Rubenstein. He refers to this as the Einstein syndrome. "Late Talking Children" and "The Einstein Syndrome: Bright Children who Talk Late" by Thomas Sowell

81 Archimedes is said to have discovered the theory of displacement on taking a bath and making a logical leap, Marie Curie's discovery of Radium was based on Becquerel's noticing that materials containing Uranium fogged photographic plates, indicating the existence of radioactive rays; Galvin noticed that when dissected frog legs were accidentally electrified they jumped. He followed it up and discovered the basis for neurophysiology, how nerves work; Fleming noticed that his bacteria cultures were killed by opportunistic penicillin mould.

82 Quotation from Louis Pasteur

serendipitous inspiration, they had already put huge efforts into searching for knowledge and challenging existing models of the world.

It is humbling to realise, when you start to read about great thinkers, how many had to deal with the fact that they were all heretics of one sort or another. Their work challenged established models. Frequently they had to leave many of their own cherished theories scattered behind them as they forged ahead. Leonardo Da Vinci has been compared to a man waking too early in the dark, while everyone else was still asleep. Many, if not all, of the greatest scientists, explorers, composers, writers and inventors saw connections where others saw impossibilities.

Isn't it time we woke up too?

# The Paradox Of Growth

> *"All that we are is the result of what we have thought. If a man speaks or acts with an evil thought, pain follows him. If a man speaks or acts with a pure thought, happiness follows him, like a shadow that never leaves him."*
> **Buddha**

## The Uncertainty Paradox

Uncertainty is the willingness to change and to learn what experience teaches us.

We all find that in life change is the only certainty. If you bind yourself too tightly to any certainty, you may find that, as that certainty crumbles like an eggshell, you will crumble with it. You must unbind yourself from certainty.

The paradox is that in order to grow, we need to have solid foundations. We need to be true to our convictions even as they change. At any given time we need to be peeling back the skin of truth to expose its structure.

Quite often it feels like sawing away at the branch we are sitting on, so why not leave it alone?

## Pruning Knowledge

If you have ever gardened you may remember the first time you came across pruning. It seemed impossible, once you had taken your pruning shears to your favourite plant or tree, that it would ever grow back. But it did grow back and it grew back stronger for not having its life force wasted on doomed suckers or dead end branches.

The tree of knowledge must be pruned so that it can grow. It is a tall tree so we must climb up and sit on it in order to get at the branches that need attention.

You are uncertain whether a new branch will grow in time for you to leap as the old one falls away but so far, so good. So far we are still breathing and the world keeps turning despite the fear-mongers and doom merchants who beseech us to resist all and any change.

## Example

When you first learn about atoms, it is very helpful to visualise electrons as balls orbiting a cluster of protons and neutrons. This gives you a model for understanding that these things exist. It gives you a way to model their behaviour.

As you go on using this image to support your learning, you learn that electrons are more like a cloud than a small planet. You are amazed to learn that they are a probability field. They do not so much orbit as promise to be there if you need them.

You have many choices at this point.

1. You can throw up your hands in despair and declare that this has been the worst let down since Chriscringlegate[83].

2. You can rail against the teacher who dumbed it down in the first place and insist that you will not trust a teacher ever again.

3. You can insist that those pesky scientists should leave well enough alone. The solar system analogy, upon which you had built your famous theory that we are all living on a molecule in a giant's fingernail, has been ruined.

4. You can cling to your very clean, certain model of things and in so doing, preclude yourself forever from the amazing things we are finding out about the subatomic world and the nature of reality.

5. You can snuggle up to your iphone and claim that quantum physics has nothing to do with real life.

---

83 The Santa Delusion

6. OR you can marvel at the phoenix of experience[84] born in the flames of a billion theories and emerging from the shell of a billion cracked certainties.

## The Ladder Of Life

This is real faith. Climbing a ladder constructed in the mind. Treating it as solid but knowing that as you pass upwards the rungs you climb will disappear into mist. Each rung did enough to raise you to the next rung. You must keep climbing though because lingering on a rung too long, treating it as the destination will result in a tumble into the abyss. This abyss is deep and littered with bodies, victims of ignorance, bigotry and fear.

> As I write this I imagine climbing out of a deep cleft between two rocks. The sun is moving across the sky and illuminating the section of the ladder I am on. Behind me the rungs that were bright, warm and safe are freezing and becoming unsafe in the shadow that climbs behind me.
>
> Ahead there is warmth and an expanding view as I get higher. I am hungry for more. My muscles, warmed by the sun, are singing to me and climbing is pleasure itself. I don't know if I will ever reach the top but I have stopped caring. I am the climb and the view is more than I could have imagined it could ever be. Each rung I climb brings fresh wonder, changing the nature of the world I think I am in.
>
> Each step is an iteration moving me steadily into new viewing positions.
>
> Around me I see the parachutes of those who chose to jump back down to replay the journey so far. Some of them are jumping from ahead so I know that there are still wonders to come. Some have jumped way below me before they got to even this height and I am puzzled why they chose to miss the wonders I have seen.

---

84  It seems to me that progress is not possible without the combustion of old certainties.

> Some ladders are vibrating but the climber is out of sight ahead of me climbing confidently, inspiringly; with no intention of ever using the parachute strapped to their back.

## Balance Is A Process Not A State.

We are human. We are much more than machines or cogs in a machine. We need to embrace our humanity and our ability to navigate uncertainty while at the same time creating tiny islands of certainty to do it from. Our ability springs from the ability to doubt. We progress by doubting the very ground we stand on and the force of gravity that holds us there. We stand shoulder to shoulder with Descartes and doubt the very fact of our existence. These doubts have resulted in our ability to view images of the far reaches of the universe. They have inspired us to discover the neuroscience to reach for the next rung. Each new rung in the ladder creates new doubts that will make us choose to create better models to doubt and so keep climbing. We need to apply the same genius to our workplaces and our societies.

Reality is in flux. Certainty creates uncertainty. Uncertainty leads to certainty. It cycles. Answers reveal questions. Questions demand answers. Nature balances. Balance creates consequences. Consequences create change and the need to rebalance.

## Have You Forgotten Project Management?

No. Project management is the art of balance. The project manager is the leader. The point of equilibrium. The leader stands at the centre of the see-saw and is versatile enough to create an island of certainty by understanding both ends. The leader challenges the certainties. The leader can be the creative paradox[85] of growth.

---

85 Gordon Mackenzie actually created this title for himself at Hallmark cards.

# The Phoenix Of Experience

*"Science adjusts its views
based on what's observed.
Faith is the denial of observation
so that belief can be preserved."*
**Tim Minchin**

## Personal Invention

When we want to filter information, we need to consider our filters carefully. We need to control them or they will control us. Many people hear the word filtering and see it as an opportunity to filter out anything uncomfortable, new or that will require a change of opinion.

Filtering works best when it is based on the legislative and executive mind working together. We want the filtered information we carry with us to have been chosen from as complete a set of representative data[86] as possible.

We need to have filters which are very deliberately inclusive. Our filters should reflect the deliberate and educated choices we make in our life. The more choices we make based on thoughtful analysis, the wider the net we can cast for relevant information. This requires an awareness of consequences.

## Personal Reinvention

Is there something for which you have a passion and know more about than the average person? Whether it is music, food, travel, art, football, books or something else, you have both the ability to make finer distinctions and to draw on a much wider palette of choice.

To get these abilities you had to learn about all the things connected with your interest. You had to learn to distinguish

---

[86] It is interesting to note that this applies especially well in understanding the importance of data in Information Technology.

what made them unique, what they are grouped with and what connected them. You had to learn to identify, distinguish and group. You had to filter, intersect and connect.

This often meant revisiting early likes and dislikes with the light of new skills and knowledge.

Many people hate some food or drink when they taste it first. As their palate matures with exposure and experience, they can often find that something that they initially rejected has become a favourite.

Whichever of your senses this passion appeals to, the expertise grew from repetition and sensitivity to feedback.

## Successful Projects Iterate

When working on projects or programs it is important to remember the lessons we have learnt elsewhere in life. When projects take too many decisions too early on, it can lead to decisions being tied into faulty assumptions. It is our nature to be impatient for progress and it can lead us to discount what does not immediately match the model we prefer.

The solution to many problems may lie in something as yet untried or something tried and misunderstood. Because we cannot see it straight away does not mean we will not come to it as we progress. We need to ask ourselves do we really understand the feedback from our experiments and first efforts. Is there a chance that when they are re-examined in the light of experience, that they will yield more than they first appeared to?

Second guessing the answers on the basis of unquestioned prejudice[87] is a dangerous occupation.

When we do this out of well meaning ignorance, we are trusting flawed filters. No matter how far we go down the trousers of reality we will remain lost until we re-examine criteria in the light of what we have learnt.

Many of the most successful projects are those that were versatile enough in their thinking to understand that some things need to be revisited as the project progresses. Getting it right first time may not mean completing it first time.

Successful managers can travel back up the decision nexus and choose a new path. This requires an ability to admit mistakes and to learn from failure. It requires that we examine the cost to benefit ratio of ego. It requires environments where people are not punished or rewarded for local optimisations, but are concerned with the intrinsic value of the work they are doing.

## Perception Is Projection

The phrase perception is projection means that we see only what our experience and knowledge allows us to see. When we observe something, we relate it back to our model of the world. We ascribe motives and motivations to other people based on our own. This leads to a lot of overestimating or underestimating based on the reality of the observer in relation to the reality of the observed.

If we have a lot of experience, we can easily fool ourselves that others understand more than they do. If we have limited experience, that is the extent of our ability to understand

---

87 Prejudice derives from creating false connections between cause and effect. If you had only ever had three traffic accidents and the other parties were hat-wearers[†] you might find yourself with a prejudice against hat wearers and assume that all hat wearers were bad drivers.

† Insert female, black, Muslim, ginger, homosexual, gypsy, immigrant or any of the other silly prejudices we are frequently encouraged to develop.

those with more education, experience, intelligence, wisdom or skill. Therefore prophets are never accepted in their own land because we find it easier to project our parochial perceptions on to them. This is why genius is often met with a metaphorical pitchfork.

Somebody from the European middle ages would see most modern technology as witchcraft. They would **perceive** in line with their model and **project** that interpretation onto what they saw.

Cargo cult is an example. The islanders did not understand the USA's technology and their model of the world was limited to their island and their culture. Accordingly they projected this onto what they observed and concluded that it was the building of runways and the wearing of headphones that brought the planes with their valuable cargo. They built coconut headphones and cut runways in the jungle and patiently waited for the cargo to arrive[88].

People project their perceptions onto animals all the time. We see animals doing things and we anthropomorphise. We think the animals have human intelligence and motivations and therefore we interpret some behaviour as human. We project our perceptions onto them. People who mess about with wild animals, projecting their perception onto them, without understanding the actual animal model, are regularly injured and killed. People who really do understand the animal models and who do not project human motivations onto them can seem to achieve amazing feats of communication with animals[89].

The same is true of everything we do. We constantly try to fit old solutions onto new problems by projecting the understanding of the old problem onto the new one. We do it

---

[88] http://news.bbc.co.uk/2/hi/asia-pacific/6370991.stm
http://www.smithsonianmag.com/people-places/john.html
[89] It is worth looking up people, like Monty Roberts and Cesar Millan, who have studied the animal's model and really understand animal behaviour and educate horse and dog owners: www.montyroberts.com, www.cesarsway.com.

with people, guessing at motivations based on some very tainted perceptions.

This is the flip side of connecting. The ability to make amazing breakthrough connections also allows us to make spurious and dangerous connections.

## Update The Projector

Have you ever watched a favourite film or read a book a second time? Have you noticed that with the knowledge of how things are going to progress, you pick up lots of things you missed the first time? Have you ever revisited some area of your life in which you have been steadily gathering skills, wisdom and hindsight? Have you also been amazed at how something, that once looked next to impossible, is now easy?

Have you ever engaged in learning something but considered your old way was better and easier than your teacher's recommendations? Have you been able to admit later that when you tried the teacher's way, a whole new universe of perception opened up? Did retrospect show you that the teacher was preparing you to step up to a new level? Did you realise that while the new way looked insane on your old level, on the new level your old way would have catastrophically limited you? Did you admit to yourself that any resistance to knowledge delayed the inevitable?

Are you proud to remember the grit it took to admit that your teacher or coach knew a thing or three and to accept guidance? Can you now do something of which you are proud, just because you embraced some information or technique that you once thought irrelevant?

This comes about because you updated your perception. We update our projector, the one that filters reality, by updating our perceptions incrementally.

Anything worthwhile elicits these step changes[90].

It all comes back to learning. We have an amazing capacity to learn, but the first step is to let go of illusions of permanence and security while questioning certainty.

## Incremental Cycles Breed Feedback And Chaos

As I filtered philosophy, science, art, management, computers, relationships, sport, business and chess while researching for this work, I kept seeing something in the corner of my mind. It kept coming back. It was insistent. I could not make it out as clearly at first. I knew I needed to heed and filter out the noise. That became the purpose of this book. Realisation. Recursion. Chaos. Feedback.

---

[90] See the discussion on the learning mechanism in volume one. Learning something new destabilises what we think we know. We appear to get worse before we get better. When we have integrated the new learning we improve to a new level then plateau and the cycle begins again.

### *Tautological feedback*

This seems to be more and more the way we understand the universe around us. It contains everything and everything in it is a perfect hologram of it. Anything that is, must be. Einstein believed that the universe does not play dice. Nevertheless it seems to love perfect tautological feedback.

- Things are the way they are because that is the way they are.
- If things were different, they would be some other way.
- Everything is always changing to achieve balance[91].
- We need chaos in order to preserve an illusion of order.
- We know that consciousness is possible because we are conscious.

### *Recursive feedback mechanisms*

- Reality is a recursive feedback mechanism.
- Evolution is a recursive feedback mechanism.
- Consciousness is a recursive feedback mechanism.

### *Responding to stimulus*

The really, and I mean REALLY, interesting thing about all this is that, unlike machines, we seem to be able to change the way we respond to stimulus and we seem to be able to initiate stimulus. This, in the vernacular, is awesome.

### *Avoiding prejudice*

Prejudice derives from creating spurious connections between cause and effect based on untested assumptions. Perception is

---

91 Entropy is discussed in Volume one

limited by maintaining those confused connections between cause and effect. Knowledge and awareness is expanded through challenging prejudice in order to more correctly attribute effect to cause.

Let me repeat that, **_prejudice derives from creating spurious connections between cause and effect_**. This has led to many misguided attempts to remove variety of response and flexibility in the name of efficiency, discipline and conformity.

When we untangle cause and effect and become more sensitive to tracing the arc of feedback, we learn from our mistakes. We learn that not all mistakes were mistakes and that most mistakes provided essential information.

### *Helpful techniques*

There are techniques in Agile and NLP which depend on this variety of action and careful attention to feedback. What has been instinctive about these is being justified by results and by developments in the study of the brain.

They attempt to attribute effect to the correct cause. They do this by:

- ✔ Shortening the feedback loop in order to disambiguate.
- ✔ Measuring and calibrating in order to identify what works.
- ✔ Repeating and refining in order to reinforce and distil.

### *Connecting to neuroscience*

If you consider what we are discovering through neuroscience about the nature of how we interact with reality, how we learn and how we communicate, these approaches are complementary and can be used as an effective framework to apply some of the breakthroughs in neuroscience to daily life, project management and interpersonal communication.

By luck, design or coincidence, the originators of these collections of techniques came up with methods which are being validated and explained by neuroscience.

These techniques sprang from an observation of what works, even when it seemed counter-intuitive or perverse. I have talked to pioneers of these approaches and they tend to be people who are very good at observing their own behaviour and the behaviour of others. They are people who are fascinated by the human factor and the workings of their own brain.

# Section Three: The Art of Knowledge

# Third Calibration

> *"Truth in philosophy means that concept and external reality correspond."*
> **Hegel**

## Models – The Basis Of Consciousness

### *Reality*

Reality is very big, very chaotic and contains an awful lot of possibilities. We experience only a small range in the spectra of light, sound and touch. In the immensity of time and space we barely exist at all. The amazing technologies we continue to develop show us that there is even more reality than we can filter directly through our conscious senses.

We are born with a set of senses and reflexes, and immediately we start to build an interface to reality. It is an interface based on experience and the feedback through those senses.

We cannot deal with reality in its entirety, there is too much of it, so we start to do the most amazing thing. This amazing thing at first looks like a fault, but as researchers into Artificial Intelligence[92] have discovered, it is at the heart of self awareness and self consciousness.

---

92 AI finds the simple things the hardest – recognising faces, navigating busy streets, understanding natural speech. Human beings find these simple. AI is good at things which have limited and understood parameters, no matter how many e.g. Chess or medical diagnosis. Human beings can find these difficult if they involve searching accurately through large amounts of data and rules.

Slides from Alison Cawsey, author of "The Essence of Artificial Intelligence": http://www.authorstream.com/Presentation/abhi_16-409553-artificial-intelligence-science-technology-ppt-powerpoint/

## Dealing with reality by proxy

We lie to ourselves. We make generalisations based on very little observation, we delete what does not fit and we distort what we need to make fit.

In other words, we get on with the business of living rather than having our mind recoil from the immensity and complexity of reality.

The result is that we all have a model of reality. It is based on our experiences. It is made up of values and beliefs. It is how we deal with reality. We interact with this model as if it were reality.

## The reversed avatar

In computing an avatar is something that represents us in virtual reality. We interact with the avatar and it interacts with the virtual reality on our behalf.

Our model of reality it is like a reversed avatar. It represents reality and the rules we expect reality to have. We interact with it and all of our dealings with reality travel through it in both directions. In computer terms it masks reality very much like the façade pattern[93].

## Mind over matter

We know that pain is not a physical sensation. Pain does not happen to our body; it happens to the model of our body that we have in our brain. The nerves convey messages of events in the limbs – we can experience pain whether those limbs are there or not[94]. This indicates that we wear a mental model of our body over our actual body. If the model is badly out of synch (e.g. some eating disorders such as anorexia), there may be trouble ahead. This is because if we hold a model long

---

93  A façade is an object that provides a simpler interface to a larger body of code http://en.wikipedia.org/wiki/Facade_pattern
94  See "The phantom limb" on page 46

enough and strongly enough, we mould our body into it until it fits – e.g. psychosomatic illnesses produce real symptoms.

The closer our model of the body is synchronised with the actual body, the more control we have over it. E.g. athletes have amazing body models because they focus on perfecting control over their physical self. Neuroscience tells us that as we practice something we increase the number of neurons mapping onto that action. The more neurons mapping onto the body the more comprehensive the model is.

If we have control over a comprehensive model that maps closely to reality, we can start to pull the strings on reality, so to speak. This is where I believe belief can alter reality. There are people who can do amazing things with their bodies and their minds. They can stop their heart, endure and ignore massive levels of pain, achieve feats of balance and agility that leave the rest of us stunned with admiration. Their belief that they can do something affects their ability because their model intersects with reality at crucial points.

## *The science of models*

Neuroplasticity has given us insights into the mechanisms we use to create, maintain and manipulate our model.

Epigenetics has given us insights into the importance of environment and how we adapt to our environment by becoming a physical mirror of our model.

## *Root concept of models*

This is the root concept in the context of these books. It is a metaphor that describes a metaphor. All perception is projection of a model. We have the ability to analyse our model and to make limitless adjustments.

## Q&A

Here are some things to ponder. They are questions I posed to myself on your behalf and I have given you my answers. I would like to think that you would have your own set of answers to these questions. It might be fun to have a go at answering them before reading my comments:

**Q.** Are there any accurate models?

**Q.** Other than your relative position to some community/society, how would you know how accurate a model is?

**Q.** What sorts of flaws might models have?

**Q.** What if flaws are internally consistent?

**Q.** What if flaws are not reconcilable with the evidence?

**Q.** Where do monsters and gods come from?

**Q.** How do models fit in with a theory of mind[95]?

---

95 See the section "Theory of mind" on page 190

**Q.** Are there any accurate models?

**A.** Not in the sense of absolute truth. That changes too much and too frequently. Static models are always losing ground against reality as it moves on. Self aware models are becoming more accurate. They are acknowledged to be models and therefore lead to learning states.

**Q.** Other than your relative position to some community/society, how would you know how accurate a model is?

**A.** Many people find affirmation in groups of like minded people with similar models – communities that now include online groups of like minded individuals (some are philosophers and some are paedophiles). I would look for inconsistencies, especially those inconsistencies rationalised, excused and relegated to the level of secondary gain[96]. There are always clues in the language being used. The unconscious mind is very truthful, even in the company of the mob. These are valuable clues.

**Q.** What sorts of flaws might models have?

**A.** Inconsistencies, over dependence on semantics, unreasonable limits on the self and others, unreasonable expectations of the self and others, limited frames of reference, denial of evidence, premature conclusions and other forms of deletion, generalisation and distortion. It is not so much flawed as misguided. Most models work and as such are all really all quite remarkable. The more incomplete a functioning model is, the more resourceful the person maintaining it has to be. Incomplete but functional models tend to trap people in cycles of unproductive and unnecessary behaviour. Resource gets sapped

---

96 See section "Secondary gain" on page 182

maintaining the model rather than reaping the benefits of all those skills.

**Q.** What if flaws are internally consistent?

**A.** They tend to be. That is the power of the mind. We hide distort and generalise things on ourselves until our models appear to be internally consistent.

**Q.** What if flaws are not reconcilable with the evidence?

**A.** Well, pretty much the same answer as the last question. People tend to distort the evidence, deny it and deliberately tie it up in semantics so that it is muffled. A popular strategy is to shoot the messenger. When you can't beat the evidence or the logic, you attack the person delivering it and make a complex equivalence between their flaws and the evidence or logic they are presenting. Public debate is full of this, equating the strength of a logical argument with the ability of its champions to argue. It is always easy to find someone who argues any point badly and then you need look no further. Whether this deletion, distortion and generalisation are deliberate or not is a much deeper question that goes to the heart of free will.

**Q.** Where do monsters and gods come from?

**A.** These are the last refuge of the fearful. When we realise our model is not explaining what we see, we create a deus ex machina to take the slack. Gods and monsters fill the gaps like expanding foam. This seems to apply to alls sorts of things. We believe in silver bullets, one-size-fits-all answers and methodologies, terrorists and reds under the bed. We are outstanding at rationalising fantasy. It really is a skill borne of our evolutionary need to ignore the sky.

**Q.** How do models fit in with a theory of mind[97]?

---

97  Theory of mind - http://www.youtube.com/watch?v=XDtjLSa50uk

**A.** A large part of communication is to infer what the other person is thinking. In the theory of mind, it is only when we start to imagine things from the other persons point of view that meaningful communication can take place. To imagine things from the other person's point of view we need to be able to model them. If we realise that these are just disposable models and we let them go, on the basis of feedback from the real person, then there is even better communication.

# Useful Concepts

*"It is a mistake to look too far ahead.
Only one link in the chain of destiny can be handled at a time."*
**Winston Churchill**

## Introduction

At the outset of this section it might be useful to examine some concepts that might be helpful.

## Stealing Fire From The Gods

> The archetypal tale of knowledge is that of Prometheus who stole fire from the gods for the benefit of humankind. Prometheus is pro+metheus and means forethought. In Greek mythology he was the creator and friend of humanity. His brother Epimetheus – epi+metheus – hindsight or afterthought – was the creator of the natural world. Prometheus and Epimetheus were the champions of mankind. Prometheus is said to have stood up to the gods and provided the human race with fire, mathematics, art and architecture. For refusing to bend to their will he suffered the wrath of the gods by being chained to a rock having his liver endlessly pecked out until Heracles eventually freed him. When Pandora opened that famous box Prometheus chose to give the human race hope rather than the ability to see the future.
>
> Knowledge is still the child of foresight and hindsight; a priori and a posteriori; induction and deduction. It loves them all.

## Secondary Gain

Secondary Gain is an important concept in understanding models. In medicine there are three levels of gain proposed to describe the underlying psychological motivations patients may have for presenting symptoms. These are primary, secondary and tertiary.

Primary gain is described as when emotional conflict is transformed into a disease or symptom. It is generally considered to be a defence mechanism. Some people become blind, deaf or unable to speak after witnessing traumatic events.

Secondary gain is similar except that there are interpersonal or societal advantages to displaying the disease or symptoms.

Tertiary gain is when the gain is for a third party. It may be that it is a sort of proxy secondary gain.

When we talk of secondary gain in a general context we are referring to all of these types of transferred gain if they hide the real reasons we are doing something. They may be conscious, but most often they have become unconscious and they may even be things we no longer desire but that have become part of our general strategy. It is not unusual for someone to feel a great sense of freedom when they uncover a secondary gain that has been dominating their behaviour and which, once exposed, can be discarded or updated.

Approaches like Agile and NLP are good at helping us to promote secondary gains to conscious goals so that we can deal with them rationally and prioritise them properly.

# Epigenetics

### *The basic theory of Epigenetics*

Epigenetics[98] is the study of gene activity, other than changes in the genetic code, that are inherited by offspring. The mechanism is believed to be the epigenome which is cellular material that sits on top of the genome. These are marks that tell your genes to switch on or off and they control the intensity of their expression. This allows environmental factors such as diet, stress and prenatal nutrition to make an imprint on genes that can be passed from one generation to the next.

It seems that behaviour can affect the epigenetic markers on top of DNA causing genes to express themselves more strongly or more weakly. E.g. behaviour such as overeating could cause the obesity genes to express strongly and the longevity genes to express weakly.

These markers can be passed on directly to the next generation offspring. It is important to

---

[98] For an easy but comprehensive introduction try *"The Genius in All of Us: Why Everything You've Been Told About Genetics, Talent and IQ Is Wrong"* by David Shenk

understand that this is different from the mechanism involved in Natural Selection. It is helpful to think of the genome as the hardware and the epigene as the software. Epigenetics does not change the genome, just the immediate programming. Epigenetics does not change DNA. It is a biological response to an environmental stressor. If you remove the stressor the epigenetic marks will fade and the DNA will revert to its natural programming over time.

It seems that the choice between nature and nurture has been overruled. We are determined by our physical and mental environment. As human beings we have choices about what that can be.

### *Lamarckian Evolution*

Jean-Baptiste Lamarck (1744–1829) was a French Soldier and Naturalist who, like Charles Darwin (1809–1882), believed that life changes gradually over time to adapt to its environment, that all organisms are related and that evolution creates many complex organisms out of fewer simpler organisms. He presented his *Theory of Inheritance of Acquired Characteristics* in 1801 – fifty eight years before Darwin's first book on Natural Selection.

Lamarck believed that an organism changes during its life to adapt to its environment (e.g. he believed that giraffes stretching for food caused their long necks) and that that this trait is passed on to its offspring.

Darwin suggested that natural selection is a result of random mutations which give advantages that make some organisms more successful than others.

The evidence has supported Darwin's elegant Natural Selection and it is generally accepted as a sound theory, particularly as it was followed by Gregor Mendel's work on genetic inheritance.

Lamarck's theories had been dismissed as incompatible with Natural Selection but with recent research into epigenetics suggesting the inheritance of behavioural traits from the previous generation through gene switch activation, it appears that we may be able to accept more aspects of Lamarckian evolution alongside those of Darwinian evolution than had been previously though possible.

## Comment

It seems to me that the modern view of Lamarckain evolution viewed through the prism of epigenetics strengthens rather than weakens Darwin's theory. It strikes me as another case of *either-or* muddying the water and obfuscating the discoveries of a scientist because of some of his personal limiting beliefs (he believed that evolution was planned) and incomplete, limited or misunderstood data (some of the experiments to disprove Lamarckian Inheritance were of the kind that showed that physical fitness was not inherited and that a dog's clipped ears were not passed on to its offspring). We need to be so careful of not throwing the baby out with the bathwater, even in apparently clear-cut cases. New discoveries can reveal new dimensions and new perspectives that allow us to make connections between previously incompatible models.

## Mirror Neurons

In the early 1990's an Italian researcher, Giacomo Rizzolatti, who was reaching for his lunch, noticed that the neurons in a monkey he was studying, fired in its pre-motor cortex as if it were the monkey reaching out for the food. The monkey had not moved. Watching the scientist reaching for his lunch was enough to trigger the same brain activity as the actual action would.

This lead to the discovery of a new class of brain cells: mirror neurons. They fire not only when we perform an action but also when we see someone else perform an action.

If you see someone else hurt themselves and it makes you wince, your mirror neurons are working. If someone smiles at you and it makes you smile back with a feeling that life is good, that too is your mirror neurons doing their job.

Before their discovery, it was believed that empathy was a logical thought process. Mirror neurons suggest that we have a built-in ability to understand and empathise with others and that it is automatic. This empathy seems to be part of the learning process. Mirror neurons are probably significant in the development of language and they are certainly necessary to allow us to read facial expressions.

It appears that mirror neurons allow us to build models of other people in our mind in the form of simulations of the intent and emotional content of their actions. They suggest that we understand others primarily by our feelings.

The existence of these fascinating and pervasive brain cells is giving us tantalising clues about our social behaviours and interactions. Some scientists and neurologists think that mirror neurons provide valuable clues about the nature of autism, schizophrenia and other conditions whose symptoms present as damage to social skills. It may be that people with these conditions may have damage to the mirror neuron system.

It seems that our ability to empathise rests on the ability to model that these special cells confer on us. Empathy may be the basis of civilisation, art and culture. It may be one of the basic tools that enables a type of Lamarckian[99] evolution and thus leads us back to epigenetics and the consequences of actions and choices.

V.P.Ramachandran calls them Gandhi neurons and believes them to be at the boundary of science and the humanities: the point at which science begins to overlap with ethics and morality.

His studies suggest that it is only the feedback from within our own system that reassures our brain that something we see happening to another person is not happening to us. Feedback prevents us from empathising totally with other people. He has observed that, if when our own limb is anaesthetised, and we see someone else being touched on their corresponding limb, in the absence of feedback, we experience the touch as a touch to ourselves.

---
99 See "Lamarckian Evolution" on page 185

This brings up some fascinating questions about the nature of consciousness and the self. It poses existential paradoxes that challenge many of our models of acceptable behaviour. Empathy and mirroring are impressive traits but in examining them we have realised that the truly remarkable thing is that we have the ability differentiate and filter what we feel from what others feel.

## Theory Of Mind

A theory of mind[100] is the presumption that other people have minds and cognitive processes. As Descartes pointed out, we can only be sure of the existence of our own mind, so we only ever have a theory that someone else might and that it resembles our own in any way.

At about the age of 3 or 4, children start to recognise that other people have beliefs, motivations and cognitive processes. This knowledge allows them to imagine what people might be thinking and to correlate that with change.

There are a number of experiments based on the ***false belief*** tasks. In these experiments children are given a little bit of privileged information that allows

---

[100] The seminal paper on this was by Premark and Woodruff in 1978 entitled "Does a chimpanzee have a theory of mind?".

them to assess the probable behaviour of someone else who does not have this information. The information requires them to either anticipate the other person's cognitive process or to rely solely on their own understanding of the situation.

For instance a researcher may show the child, Johnny, and another person, Sally, a ball being put into a red bag. Sally leaves and the researcher moves the ball to the blue bag.

When Sally comes back Johnny is asked where Sally will look for the ball. Before the age of 3 or 4 Johnny will point at the blue bag where he knows the ball is. At this stage he is not allowing for Sally's cognitive process or "mind". At 3 or 4 Johnny will know that Sally last saw the ball in the red bag. He will realise that from Sally's perspective the ball is still in the red bag and will indicate that this is where she will look for it.

There are a number of variations on the experiment but they all seem to indicate that we reach a stage of development where we definitely start to understand that other people may have a different model of the world and that we can estimate what they might be thinking.

Research into autism also points to a theory of mind being necessary to understand and predict other people's behaviour. It appears that a theory of mind is necessary for successful interaction with other human beings and for the ability to react adaptively to a changing environment.

This is a formidable ability that allows all sorts of communication, community and collaboration. It is also a liability when we forget that we are guessing and we think we really know what other people are thinking, based on generalised, distorted and incomplete information.

## Boundaries And Constraints

In this book I chose to make an important distinction between boundaries and constraints. Constraints stop you from doing things. Boundaries create spaces for you to do things. Historians used to believe that the 120km long Hadrian's Wall between Scotland and England was built by the Romans as a defensive measure against the Scots – a constraint. Many historians now believe it was not built to keep the Scottish out of the Roman Empire at all but as a statement to declare the edge of the Roman Empire and Roman responsibility. They believe it was a visible reflection of the power of Rome and that it was built to say this is the edge of the Roman Empire. On this side you are protected by the power of Rome – a boundary.

## Response Ability

*"Reparation was their main excursion into the economic field, and they settled it from every point of view except that of the economic future of the States whose destiny they were handling."*
**John Meynard Keynes regarding the Treaty of Versailles**

### *The mirror of versatility*

In various social engineering projects, nanny states have tried to regulate everything, only to find that as risk disappears, so does innovation and creativity. The response to making something illegal is to increase its popularity. The response to censorship is an increase in curiosity.

The safer we try to make our children, the more we kill their spirit. Disinfecting everything lowers their immune response and makes them more susceptible to illness.

Risk is the parent of skill. Obsessive control leads to havoc and chaos leads to balance.

Safer cars can lead to more accidents rather than fewer. Studies suggest that the best way to improve road safety would be to place a large spike in the centre of the steering wheel. This is called Risk Compensation or Risk Homoeostasis. People perceive more safety and are willing to push up the risk factor to keep the fatalities constant[101].

The more people feel controlled, the more they are likely to rebel against it.

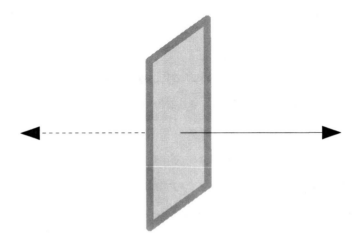

Powerful tools and powerful abilities are double sided. Knowledge itself can be used for great good or great harm. Weapons and tools can be made from the same technology. Dynamite was developed to be used as a construction tool before it was used to kill and maim. The plastic surgery used to help burns victims was born out of the need created by the Second World War. There is a mirror of versatility.

The sharper the blade, the less restricted the tool, the more useful it is and the more potential danger it

---

[101] "Buckle up your seatbelt and behave" By William Ecenbarger
- *Smithsonian* magazine, April 2009

brings. If you can't find a misuse for something you probably can't find a use for it either. This is the price we pay. Safety is an illusion. The more latitude you give people to fail, the more likely they are to succeed.

For something to be useful it also seems to need the versatility to be harmful.

Picture it as a type of plasticity: as something is pulled from neutrality to effectiveness it casts an equal reflection of potential harm on one side and potential benefit on the other. As it is extended into effectiveness in any direction a mirror image is being stretched in the opposite direction. Intent is the controlling factor. Intent cannot be legislated. Responsibility is the parent of intent. Every ability creates a response ability. It is your choice as to how you use it.

## *A Priori* Tea

The 18th century British industrial revolution owes its success to tea. Apparently tea contains substances that helped the British worker to avoid illness. This resulted in a more available and healthy workforce and people being able to live in big cities without succumbing to devastating diseases. If this is true, the much maligned tea break was an essential ingredient of the industrial revolution.

> Tea was originally an herbal remedy in China because of its antibacterial properties. When it was introduced to England it was closely examined. Experiments showed that frog legs in water putrefied while those in tea did not.
>
> Around 1740 the infant mortality rate halved across rural, urban and class divides in the UK. The rise in popularity of tea coincides with improving health which made breast milk healthier. It cannot be explained by advances in medical science that were yet to happen, or improvements in sanitation and hygiene that were still in the future. Historical records show a reduction in the incidence of water-borne diseases in that period.
>
> Tea is made with boiling the water and according to Professor Alan McFarlane[102], Professor of Anthropological Science at Kings College Cambridge, the tannin in tea is a powerful antiseptic. It probably helped people avoid water-borne diseases such as dysentery, cholera and typhoid.
>
> Improvements in health allowed the growth of large cities. Tea drunk by the workers in those cities contains a stimulant and provides energy with the milk and sugar taken with it.
>
> All this adds up to a convincing argument for the effect of the tea break on the industrial revolution.

The planning of any project or endeavour is the measurement of the change that is happening to it. If you do not notice what is changing, you are measuring the wrong thing. The project will become difficult as it tries to tell you what is really going on.

Priority gives us the most basic tool of planning. We must know, or have a way of finding out, what is important. We must start with some knowledge,

---

[102] Professor McFarlane's site
http://www.alanmacfarlane.com/savage/tea.html

something we can hold as true, in order to measure change.

Assumptions are made on *a priori* knowledge. *A priori* knowledge is transformed into *a posteriori* by the doing of something and the awareness of distinctions and shades of change as it happens.

Imagine a factory owner at the time of the industrial revolution who has seen that longer hours worked are resulting in more productivity and more profit. It would not be unreasonable to imagine that he would want to dispense with anything that was keeping his workers from their work.

In fact between 1741 and 1820 industrialists, landowners and clerics tried to put a stop to the tea break. Workers took a stand, even before there were trades unions, and the tea break remained an institution.

*A priori*, that which we know in advance, allows us to induce much more. It allows us to know what the priority is. Like a dragon eating its own tail – in a basic recursive feedback loop – this *a posteriori* becomes the *a priori* for the next revolution.

Tea drinking allowed people to work longer hours. Longer hours gave the opportunity for more tea breaks. Back then nobody was making the connection between tea and a healthier workforce. The research that provided this link came in the 20th century.

The point is not to advocate tea drinking, but to highlight the recursive nature of very important things and the danger of creating incomplete models of cause and effect.

Where would the industrial revolution be if the industrialists of the 18th century had been successful in abolishing the tea break which seemed to them to be an unnecessary cost? How many companies and governments are currently up to their armpits in cost cutting as policy?

To put it more simply, like Archimedes, you make your best guess and stand on that spot from which you think you can move the world and then you apply your lever. Once applied, you observe the effects as honestly as you can and you adjust your position until you inhabit the sweet spot.

You can only do this by knowing where the fulcrum is. You can only know what your real priorities are by shortening the feedback loop and loosening limited visions which were based on limiting beliefs and limited thinking.

You must refresh the *a priori* to establish a true priority, even if it turns out to be the tea break.

## Nominalisation And Denominalisation

To nominalise is to turn a verb into a noun and to denominalise is to turn a noun back into a verb. Nominalisation can cause us to confuse states with processes. It is important to distinguish between those things with which we can interact and those things that define relationships. In "The Structure of Magic", the original NLP book, Bandler and Grinder suggest that nominalisations and denominalisations cause confusion and they propose ways to identify and deal with them. They suggest the wheelbarrow test. If you cannot put it in a wheelbarrow and wheel it away, then it is a relationship or process rather than something you can interact with or measure. (You might have to imagine some very big wheelbarrows for some things). Can you put love/goodness/happiness/change into a wheelbarrow? If not they might be usefully denominalised. Similarly if you can say "ongoing" before a noun, it may also be a process. Ongoing love but not ongoing bricks. By the preceding definition is reality a nominalisation or a denominalisation? Is reality, like love, happiness, change and the other things that are ongoing and escape the wheelbarrow, a process rather than a state? Is it something constantly changing that we can work at, that we can accept or ignore; or is it a permanent condition? If so, what does that tell us about our models?

## Strange Loops

There are abstract loops found in almost all models of reality including mathematics, art and music[103].

The easiest way to illustrate it is possibly with M.C. Escher's art. "The Waterfall"[104] is a good example. It is a picture of a waterfall that starts and ends at the same place. Escher manipulates perspective and his knowledge of how our mind translates two dimensional representations into the three dimensional things they represent. This is obviously done with deep knowledge of perspective, but the underlying idea is fascinating and mesmerising. The important thing to remember is that the brain makes the leap between the two dimensional reality of the map and the three dimensional reality of a waterfall. It points out that perspective is in our mind, not on the paper. This is a subtle teaching.

Explaining what happens in music is more complex. Indeed it is tempting at this point to

---

103 Douglas Hofstadter talks about them in his book "Gödel, Escher and Bach".
104 http://www.mcescher.com/Gallery/recogn-bmp/LW439.jpg

abandon all else and just devote the rest of the book to the neuroscience of music but it has already been done[105]. Bach's music is mathematically compelling and can be used as an example of auditory strange loops[106].

There are various theories and very well argued works about how the brain perceives music. It is a minefield through whose semantics I have no intention of gaily skipping.

The colours I will pin to the musical mast are these:

- Music exists only in the brain and is dependent on a theory of mind. We have to assume that the composer or musician has an idea or emotion they are trying to communicate through these sounds[107]. Van Morrison called it the inarticulate speech of the heart in his 1983 album based on the idea that emotional articulation is not dependent on words. It seems that one kind of articulation may be inversely proportional to the other.

---

105 Oliver Sacks M.D., professor of neurology and psychiatry, author of "The Man Who Mistook His Wife For A Hat" and other seminal works on neurology has already done this with his 2007 book "Musicophilia: Tales of Music and the Brain" and a film "The Music Never Stopped" based on his essay "The Last Hippie". Daniel J Levitin's 2006 book "This is Your Brain on Music" and his documentary "The Musical Brain" explore the same fascinating territory.
106 http://www.youtube.com/watch?v=A41CITk85jk this illustrates how a piece of Bach appears to spiral up the scale but it always ends up back at C minor
107 See the note on Chopin and Abstract Music on page 330.

- Talking about listening to music can vastly overcomplicate something that many people believe is intrinsically entwined with their humanity. I frequently find listening to music high on my list of reasons to exist. Depending on whom you listen to, music either drove evolution, evolved side by side with us or is an incidental by product of evolution. These arguments present too many *either-or* choices and make me doubt whether anyone really knows.
- To interpret vibrations in the air, and the electrical signals they cause our auditory equipment to send to our brain, we use memory, categorisation, anticipation and prediction. We remember the last note, anticipate the next and create relationships of tone, frequency, pitch, timbre, melody, harmony, rhythm, tempo, dynamics and form that exist only in our mind.
- Music is an architecture in time. Each note exists in its own in time frame but we connect them with our listening and create a continuum we label music.
- The subtleties you can appreciate increase with practice. Your choice of the music that you listen to determines your appreciation, your ability to make distinctions and your preferences.
- Music is more than sound. It is the brain of the composer, the brain of the listener and the laws of

physics and neuroscience combining the imaginary and the real to create strange and complex loops.
- Music relies on a sense of perspective to create an "other" relationship. Music is a nominsalisation of the perspective that adds dimensions to simple sounds that can touch us at our deepest levels and alter our deepest maps.

Elsewhere we have discussed how Gödel's mathematical ideas and Bertrand Russell's logic make use of strange loops to explain that mathematics and logic are merely maps or models of reality. They have the same limitations as a two dimensional picture trying to represent three dimensional reality. Now consider that science is telling us that the reality we model in four dimensions probably has more than eleven. Strange loops become less of an interesting phenomenon and more of an essential tool.

## Strange Loops

A strange loop is a sort of paradox whereby moving through successive hierarchies one arrives back where one started because there is some kind of "other" relationship between the hierarchies.

In art the "other" relationship is that we know those two dimensional lines represent three dimensional reality, so that is what we see. In fact once you see what is being represented it is hard to see the technique any more.

This "other" relationship is the stuff of genius. It is the root of inspiration and it is still the domain of the human mind that can connect laterally and follow through vertically. The human mind can often go through a logical wormhole to a solution without following sequential steps[108].

---

[108] Logical wormhole and vertical and lateral connection are dealt with in next chapter.

# Envisioning Creative Thinking

*"every opinion now accepted was once eccentric."*
**Bertrand Russell**

## As If Slicing Reality

Whatever way we slice it, reality presents different faces to us in different places, different times and different states of mind with a different attendant set of dependencies and responsibilities.

When we are on holiday, reality can seem so different than the unimaginably distant office and daily grind. On the Monday they get back from holiday we often hear people saying that it is as if they were never away. The holiday reality is totally divorced from the work reality. We can meet old friends after long separations and feel like we are continuing where we left off as the years in between cease to exist. Reality presents a complete and distinct façade to us in our role as parent, worker, our hobby or our social identity. Each reality can seem to be complete and singular as we inhabit it.

Every project has an operational, economic, technical and social reality. They seem at times to be mutually exclusive yet entirely necessary.

Other examples might be the different slices through reality we are presented with in our jobs, as a member of a team, a project, the wider group, our family and our sports group. They may be more orthogonal than concentric for many people.

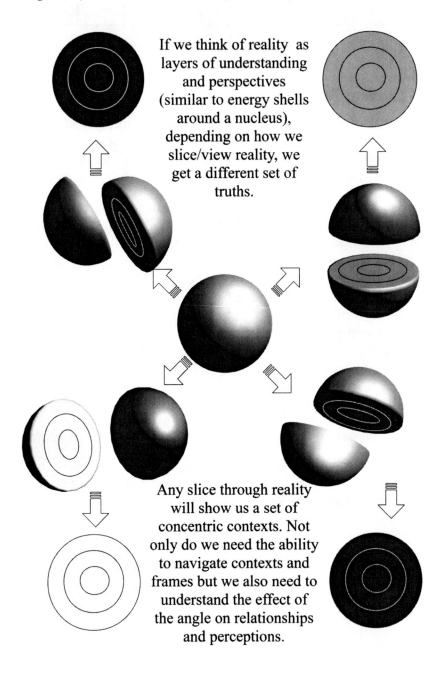

If we think of reality as layers of understanding and perspectives (similar to energy shells around a nucleus), depending on how we slice/view reality, we get a different set of truths.

Any slice through reality will show us a set of concentric contexts. Not only do we need the ability to navigate contexts and frames but we also need to understand the effect of the angle on relationships and perceptions.

This is what I mean by slicing reality: how our consciousness can present us with different, complete and internally consistent realities. Each of these realities has its own set of concentric contexts and each of these realities seem to exist orthogonally to each other in alternate universes. This orthogonality can cause us to perceive intractable problems when we try to resolve their respective essential requirements of us.

## The Slices Out Of Context

*This is the mirror process to **Connection is the mechanics of inspiration** on page 151.*

When we try to find the intersections of different pieces of reality, different aspects of our life or areas of knowledge and expertise, we may often find that we have conflicting knowledge and certainties. This leads to a state of paradox. It is the result of the assumption that all the slices we see are from the same plane. Therefore we see only the intersections on the plane we are observing – i.e. the current frame of reference.

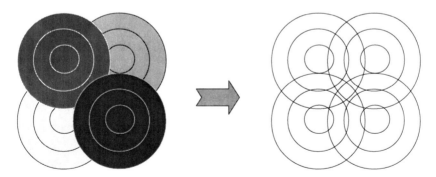

This can be complicated by the fact that we have narrowed our context as well as making false assumptions about the nature of reality.

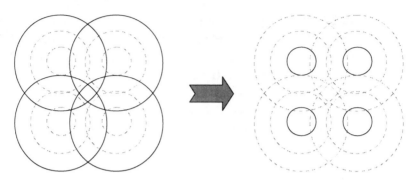

We do this with deletion, generalisation and distortion. We arrive at situations we perceive as problems because we have entangled these elements with limiting beliefs, complex equivalences and harnessed the lot to our own ego. We have deleted the connections. If this is the case we may not even see the intersections possible in this frame of reference because of our limited context.

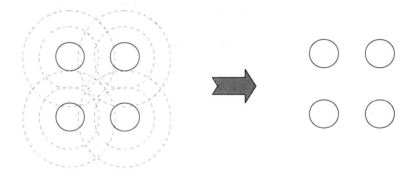

## Logical Wormholes

In physics a wormhole is a hypothetical feature of spacetime. It is theorised that, as spacetime bends according to the theory of relativity, it might be possible to take a short-cut.

If we imagine space-time as a two-dimensional sheet, the distance from A to B seems to be straightforward.

If spacetime can be folded, as it appears to be by gravity, we can find a shorter route from A to B. This theoretical short-cut is termed a gravity wormhole. If space and time, the building blocks of reality, have these qualities, is it really strange that our daily lives sometimes require us to engage in a spot of creative thinking to resolve issues?

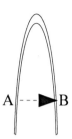

Now imagine solutions as the connection between concepts, between different areas of knowledge and between resources or expertise. They may appear to be inconsistent but if we assume an "other" relationship, we can start to discover strange loops in the paradoxes. At the turning point of the strange loop we can often glimpse the singularity and the pieces resolve into a possibility. This is not a process, it is intuition, insight and inspiration. They are creative states of thinking.

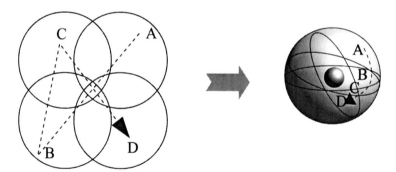

Consider Einstein contemplating the problems of light, gravity, time and space. All appeared to hold truths incompatible with each other. He collapsed these paradoxes into an idea of relativity where time is relative, light is constant and gravity is a distortion in the fabric of space.

He is attributed with having said "To raise new questions, new possibilities, to regard old problems from a new angle, requires creative imagination and marks real advance in science."

As in the case of Einstein, the mind can ride on a beam of sunlight and connect the dimensions of relative reality to the heart of a solution. From that solution the mind can lay the slices back out, side by side, and track the consequences in each context.

Creative thinking is a part of life in the information age and if we were to visualise it, this sphere is as good a metaphor as any.

There are processes for generating creative, insightful and intuitive states. My concern is that these processes – no matter how effective they may be at the start – always seem to become confused with the states they were designed to encourage. This can reach the stage where the processes are being confused with the effective states as the cause of progress. When this happens creativity, experimentation, insight and intuition are seen as overheads to the process. This creates a *process strange loop* in which the process becomes its own end (E.g. the minute your goal becomes "*to be an Agile company*", you lose the flexibility of Agile.). If we are on the lookout for strange loops as triggers, we can collapse these processes back into something resembling a useful tool.

## Vertical And Lateral Thinking

Vertical thinking means that you take each step based on the results of the last step and build a solid logical argument to get from A to D. This is fine but it is restricted. You can only get to places that can be got to from where you started and you rely entirely on the direction of your argument. If you start in the wrong place you may deny yourself many possible outcomes. It relies heavily on negation and filtering out information on the basis of the current frame of reference. If there is any mistake in context, the frame of reference or the categorisation of information, vertical thinking may deliver a sound argument but a flawed or at best limited and incomplete conclusion.

Lateral thinking[109] can increase the possible solutions. It makes use of the fluidity of context, frame of reference, categorisation and labels. The seemingly irrelevant, the surprising, the intuitive and changes of perspective are fuel to lateral thinking.

You don't have to choose between them. It is not *either-or*. They are tools. The human mind is capable of using both vertical and lateral thinking. It can connect laterally and follow through vertically. The human mind can often go through a logical wormhole to a solution without following sequential steps and then find its way back the long way with the solution in tow.

## Intuition

Tuitionem is Latin for guardianship from which we get the idea of guardianship of knowledge and of passing it on and teaching. Intuition is your inner teaching. It is making use of your accumulated learning and all of your resources. It is perceiving directly and bypassing conscious reason.

## Insight

Sight with the mind's eye. Plato considered noesis (from the Greek noein to have mental perception and noos meaning mind and thought) to be the highest form of knowledge. Philosophers through the ages have sought to understand it. Noetics is the study of those harder to define gut feelings and subjective understanding. One theory is that the unconscious mind draws on reservoirs of information and unconscious observation to predict what will happen next or to make leaps of understanding. Naturally there are some seriously crackpot

---

[109] Edward De Bono's book "Lateral Thinking" is a thorough discussion of these types of thinking.

theories out there about all this, but from Plato and Aristotle to current day philosophers and scientists the idea persists that the mind has cognitive abilities beyond those we routinely recognise.

## *Inspiration*

Comes from the word spirare to breath. It carries with it an idea of being consumed with a purpose or idea to the point where we are indistinguishable from our purpose, we inhabit it totally. It is the doorway to the state of being in flow.

## Access Genius – Reconnect The Slices Of Reality

When we want to access our creative or leadership thinking we reverse the slicing process. We can use logical wormholes to move through various frames of reference. This gives us a better perspective. We actually use the phrase that we have a new angle on something and we often use it just before we discover a solution.

*Envisioning Creative thinking* | Page 213

1. **Filter:** Cast your net of combined intuition and logic. If the pieces of the problem were to be related, at what level are they related?

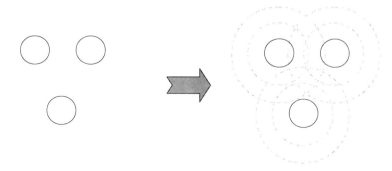

2. **Intersect:** From this level how do these ideas overlap? Where is the fulcrum?

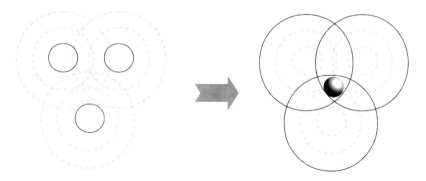

3. **Connect:** If there were to be a nucleus of all these ideas, where would it be?

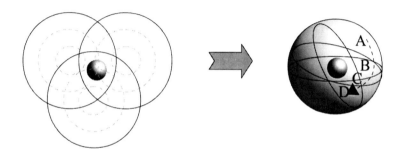

## Creative Thinking – A Conclusion

In a sense creative thinking and problem solving require you to dissociate from the problem. When we dissociate our egos, we also dissociate conditions, complex equivalences and limiting beliefs from our thinking process.

Filtering requires you to filter these things out and filter in possible solutions that may only need a touch of refinement to become working solutions. It is a non adversarial, non competitive, collaborative and ego-less approach.

All of the above can be summarised as follows: From what perspective can any situation be viewed so that it is not a problem? When I can find that perspective, I know the solution already. Then it is only a matter of refining it.

Very often these leaps into the dark turn out to be well lit. They can take you onto far more solid ground from where you will have the opportunity to apply reason and create balanced solutions.

# Feedback Loops

*"If all the economists were laid end to end, they'd never reach a conclusion."*
**George Bernard Shaw**

## Learning Through Feedback

Learning is the natural state of all human beings, perhaps of all living beings. Learning can only take place where there is some sort of feedback.

There are two main sorts of feedback loops recognised in engineering. If we recognise feedback as a universal principle, they are worth knowing:

**Positive feedback** – reinforces the event that started it until it hits a constraint. The constraint is quite often destruction.

E.g. a thermonuclear reaction (in fission – breaking atoms apart – and fusion – joining atoms together – each individual reaction can cause one or more similar reactions in a self sustaining chain. In fission, particles are released and interact with fissile material in a self sustaining chain reaction until the material is depleted. In fusion, it is dependent on temperature and pressure conditions maintained by the energy release from each fusion process.)

**Negative feedback** – balances the event which it is measuring until it reaches a predefined setting. Delayed feedback can turn a negative feedback loop into a positive one and can also end in destruction

E.g. a hot water thermostat (if there is a delay between heating and the measurement and adjustment, the heating element could overheat and burn out)

A simple feedback loop might look like the following:

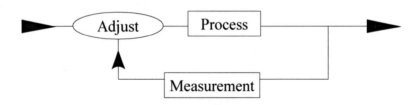

A process takes place and the effect is measured. This leads to adjustments before the process is repeated. The outcome of the loop is to reinforce some behaviour and regulate the system.

## *Reinforcement*

**Positive Reinforcement** – providing an appetitive stimulus to encourage certain behaviour.

**Negative Reinforcement** – application of an adverse stimulant to discourage certain types of behaviour.

All of the above depend on the ability to associate the effect with the cause.

### *Single cause: Single effect*

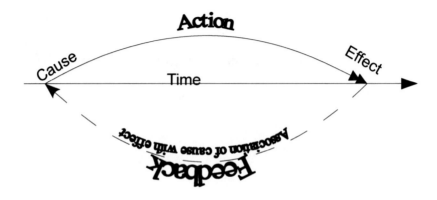

## *Feedback is learning*

This is an important part of the learning mechanism. We do something and it has an effect. We associate the cause with the effect and we make hypothesis about the world.

Babies are masters of association: "If I cry someone comes and feeds me. If I push this, it falls over. If I bite my finger it hurts."

## *Some important feedback principles*

- ✔ The closer the cause to the effect, the stronger the association.
- ✔ The bigger the ratio of effect to cause, the stronger the association,
- ✔ The more repeatable the effect: the stronger the association.
- ✔ The more singular the effect, the stronger the association.

## *Getting the best from the feedback mechanism*

- ✔ Cause and effect need to be close to each other in time.
- ✔ Effect needs to be magnified.
- ✔ The effect loop needs to be consistent and repeatable.
- ✔ Effect needs to be highlighted and identified clearly.

If only one of these rules is broken we still learn. But what do we learn?

- ✗ If a cause is distant in time from an effect we can miss the association.
- ✗ If we do not notice the effect, we do not make the association.

Page 218 | *Section Three: The Art of Knowledge*

- ✗ If we do not notice the effect every time, we do not trust the association.

- ✗ If we confuse cause and effect, then all sorts of very strange associations begin to form.

## The Cause – Effect Ecology

In the real world there is the issue of multiplicity. Nothing happens in a vacuum[110].

This happens a lot:

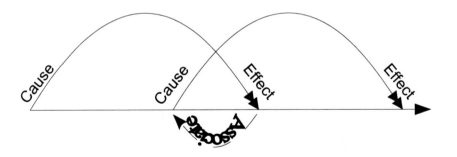

There can be a tendency to associate a consequence with the most recent action. We must learn to untangle coincidence from consequence.

---

110 (unless you're an astronaut or James Dyson).

## *Multiple cause: Multiple effect – sequential*

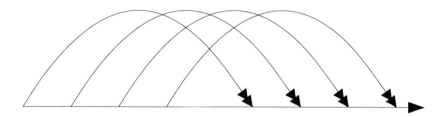

In the diagram above we have four causes. All four are initiated before there is an effect from any of them.

Which effect do we associate with which cause? Maybe we just associate the first effect with the first cause?

Is it just a matter of remembering the order and associating the effects in the order in which they were initiated?

Then what about this?

## *Multiple cause : Multiple effect – nested*

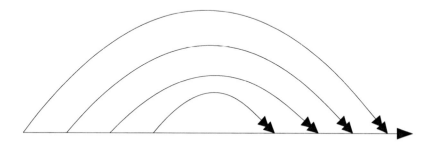

Cause and effect are nested. We have four causes all initiated before there is an effect from any of them. In this case the effect is not linear and the effects arrive in reverse order to the cause.

## Computing cause–effect metaphor

In IT there are patterns for dealing with both of these models. We can isolate those things in which we are interested. Because computers do things very quickly and we can write code very slowly, we can examine time at any speed we want to. This allows us to treat cause and effect in a way that reality rarely allows us to. We can manipulate cause and effect as if it were sequential or nested.

1. Queue or FIFO (first in first out) can deal with sequential cause and effect.

2. Stack or LIFO (last in first out) can deal with nested cause and effect

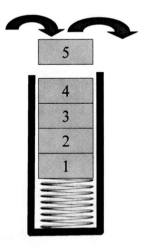

Reality has a few more variations for us to consider:

## *Multiple cause : Multiple effect – random*

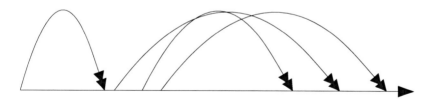

We can have causes whose effects come back in a totally random pattern. There is no discernible or predictable order to when the effects will happen with respect to the order of the causes. This is more like real life.

Consider your bank account. Most of us have cards with which we make purchases. The application of a charge to our account is not instantaneous. Some come through the same day and some take up to a week. If you are still in the habit of writing cheques, then it can take up to six months before the charge hits your account. In both cases the charges can come into the account in totally random order unrelated to the order of the purchases.

Then you have direct debits and standing orders. With a standing order the same amount comes out every month but with a direct debit the amount can be changed by the payee. Service companies prefer you to pay with direct debit. They are sometimes known to make mistakes and overcharge[111]. Although standing orders and direct debits both have payment dates, usually they get applied to your account a couple of days either side of the agreed date.

Does the current balance ever reflect the most recent purchase and how can you gauge the effect of which payments have been applied or have yet to be applied to the account?

---

[111] Dealing with this situation is enough to make anybody turn Luddite, reinstate the grand bank of the mattress and cancel all their direct debits out of sheer terror.

Balancing a current or checking account, even if you have a great memory and keep all the receipts, is a difficult job. Unless you have total recall at a very detailed level, it is hard to reconcile the fluctuations in your account with your real-time spending[112].

## *How we associate complex cause and effect using submodalities*

One theory is that we encode our experiences. We tag them with unique attributes so that we can access them again in order to associate cause and effect. NLP uses the metaphor of sensory *submodalities* to explain this.

The sensory modalities are Vision (V1), Audio (A1), Somatosensory – touch or kinaesthetic[113] (S1), Gustatory (G1) and Olfactory (O1). These relate to discrete areas of the brain. We know that these areas are split into further refinements, or *submodalities*, so that we can recognise distinctions. In vision this would be colour, distance, 3D, motion, stillness, brightness, focus etc. Sound would be loudness, pitch, proximity, tone, mono, stereo etc. Kinaesthetic would be temperature, pressure, shape, speed, location etc.

As with cause and effect, we can make these distinctions to isolate and access events. Sub-modality attributes can also be cross wired in synaesthesia to create various unique combinations.

This is a very important concept within NLP which allows both access to and manipulation of perception and learning.

---

[112] This is the same reason there is so much process around project budgets. The real time state of a company's accounts are frequently shrouded in mystery and guesswork.

[113] There is a semantic distinction between Kinaesthetic and Somatosensory as Somatosensory includes proprioception which is the sense of where we are in space. It makes balance possible. When people talk about kinaesthetic representational systems in NLP they would usually include touch, movement and proprioception.

It can be used to create helpful associations and release unhelpful, limiting or phobic associations[114].

We are particularly sensitive to repetition. Neuroscience highlights that we are very sensitive to causes and effects that tend to happen together. We instinctively and automatically cross wire causes and effects because we know that *"once is happenstance, twice is coincidence, three times is enemy action."*[115]

We know that cause and effect is much more complex than the models presented so far. Let's look at some more.

### *Single cause : Multiple effect*

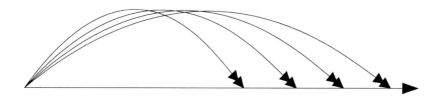

We know that when we do things that they can have a ripple effect. It is important to be aware of more than just the primary effect we are interested in. If we want to be ethical and ecological we must be aware of side effects and risks.

### *Multiple cause : Single effect*

---

114 A basic practitioners course in NLP teaches this.
115 Attributed to Ian Fleming – creator of James Bond

As adults we know that many things are the culmination of multiple causes. We know better than to associate the last thing that happened as the sole cause – at least, we should. It is surprising how many people forget this and the result is what is known professionally as "a knee jerk reaction".

## *Single/Multiple cause : Multiple/Single effect*

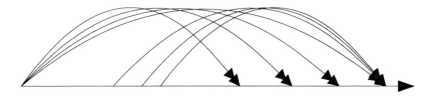

## *Cause and effect complexity within a frame*

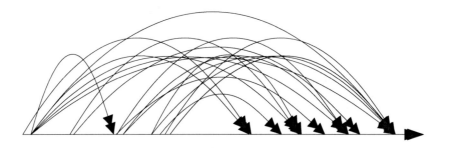

Within any frame we choose to view, the complexity of cause and effect can be difficult to unravel. Many causes and effects exist side by side. If we choose to treat any event as an isolated incident and choose to examine only those causes and effects within its time-frame, all sorts of apparently disconnected causes and effects will intrude on that reality to confuse it. We often need to frame up and out to start understanding even the most simple of interactions.

## *For a better analysis, widen the frame*

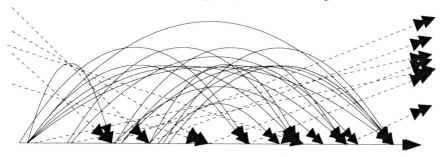

Bearing in mind that the world did not pop into existence for that event, it has got to be influenced by things reaching out of the past and it will almost certainly influence things into the future.

### *Effects are consequences*

An effect is a consequence. We have to deal with consequences all the time. We have to consider consequences all the time.

## Recognising Interference

The world is a chaotic place and it is getting more chaotic. Not only do we have to deal with the essential noise of the system but increasingly we have to deal with forces that deliberately create cross wiring in our cause/effect mechanism.

### *Politics*

To gain or hold on to power, politicians deliberately confuse cause and effect and thereby never have to answer a straight question or address the real issues across tribal lines.

### *Marketing*

To manipulate us into buying what we would otherwise just ignore as irrelevant or too costly on many scales, some salespeople, commercial companies and the advertising empire deliberately confuse cause and effect so that we are focussed on what they want to sell rather than what we really want or even need.

### The Media

Some parts of the media like to create sensationalism, pander to prejudice and fear, and push various agendas. They often deliberately confuse cause and effect to achieve this.

### Bullies

To further their own careers the ruthless, in almost all walks of life, confuse cause and effect to take credit they do not deserve and to apportion blame where it does not belong.

### It's not new

To gather power, wealth and various pleasures, totalitarian belief systems, since time immemorial, have confused cause and effect[116] in spectacular fashion. This has resulted in the abuse – mental, physical or existential – of just about everyone who has ever lived.

## Self Defence

We need to protect ourselves against the onslaught whilst preserving our ability to adapt and learn. In order to advance, and perhaps survive as a species, we must untangle and appreciate cause and effect:

---

116 Confused cause and effect:
- The crops failed because the gods are angry.
- You are sick because you are a sinner.
- That happened because you did not say the right incantations in the right place at the right time.
- You will be happy if you give us all your money.
- The earthquake was caused by civil partnerships.

- ✔ Untangling by understanding the consequences of things, by joining the dots and by opening our eyes to all the consequences and what they are consequences of.
- ✔ Appreciating by accepting the consequences of our own actions and reaching beneath the simple surface structure of wishful thinking and selectivity that hampers the learning loop and reinforces the ignorance loop.

The way to do this is to shorten the feedback loop where we can and hone our ability to detect and question generalisation, distortion and deletion. We have to develop crap detectors and keep them up to date.

## Out Of Complexity

There is something else on our side. It redresses the balance in the most unexpected of ways. It is chaos.

# Order And Chaos

*"Out of intense complexities intense simplicities emerge."*
Winston Churchill

## Butterflies And Cats

Chaos is sometimes visualised as a butterfly[117]. This is the butterfly that flaps its wings and causes a storm on the other side of the planet because of the deterministic and connected nature of reality. The air molecules it dislodges dislodge more air molecules in a cascade of cause and effect that results in tornadoes. In a chaotic system, very simple and insignificant causes can have enormous and destructive effects.

A butterfly flaps it wings in Brazil but it does not cause a storm anywhere. The reason? Chaos does not work like that at all. The butterfly was a teaching metaphor similar to Schrödinger's Cat. Neither the cat nor the butterfly really exists except as pitons in the walls of reality. The minute you start to think you can predict the effects of chaos you have locked yourself in the tree-house of truth[118], kicked away the ladder[119] and thrown out the key.

## Deterministic Universes

In a Newtonian universe everything worked like clockwork. There was a certainty that everything was predictable as long as we could discover the rules. It was safe to assume that reality was ordered and under control.

---

117 Edward Lorenz (1917-2008) mathematician, meteorologist and pioneer of chaos theory introduced the metaphor of a butterfly causing the storm in his 1972 paper "Does the Flap of a Butterfly's Wings in Brazil Set Off a Tornado in Texas?". He was talking about the unpredictability of the weather because of its sensitivity to initial conditions.
118 See "Thinking inside the box – the Treehouse of truth". Page 242
119 See "The Ladder of Life". Page - 161.

Anything that challenged this predictability or disproved it was considered to be a malign external influence. If the results were not as predicted, it was considered that there was either something wrong with the process or someone was fiddling the results. So the results were fiddled back to what they should be.

In the natural world this control is hard to detect. There is a strange mix of pattern and irregularity, predictability and unpredictability, simplicity and complexity, all living side by side.

The Newtonian universe was a surface structure. Chaos lies deeper.

## Deterministic Chaos Theory

What Chaos, the scientific term[120], says is that even within completely understood, simple deterministic systems, there will be unpredictable behaviour. Even if these systems are completely mathematically understood and there are no random events or outside influences, there will be unpredictability.

## Sensitivity To initial Conditions

Perhaps the most important thing to understand about Chaos Theory is that it is not about randomness; it is about sensitivity to initial conditions. Chaos is simplicity itself. It follows simple rules to the letter. It follows the instructions of the simple mathematical rule of self similarity and feedback. It also, deviously, honours the initial conditions. If these are misaligned, by even the smallest degree, chaos will do very interesting things.

---

[120] To get a better understanding of the science it is worth starting with the work of Alan Turing (1912-1954) on morphogenesis, Boris Belousov (1893-1970) on non-linear chemical dynamics, Benoit Mandlebrot (1924-2010) on fractal geometry and Lord Robert McCredie May (1936- ) on ecology.

In book one, we looked at a phrase being passed backward and forward through a language translator and how quickly it mutated away from the original intent because of small imprecisions in meaning in the original phrase and tiny faults in the translator. This was a little bit of chaos in action. In that example I was demonstrating how chaos can be used to test. Chaos teaches you very quickly how much you do not know about your initial conditions.

In computing and business these initial conditions can be your requirements. They can also be your process, your design, your skill set and your management approach.

## A Simple Complexity

$$Z \rightleftharpoons Z^2 + C$$

This is a feedback equation that produces the Mandelbrot[121] set. It works by iteration and recursion. The output from the equation becomes the input for the new equation.

$$Z \rightleftharpoons Z^2 + C$$

$$Z \rightleftharpoons (Z^2 + C)^2 + C$$

$$Z \rightleftharpoons ((Z^2 + C)^2 + C)^2 + C$$

and so on.

A diagram can be programmed from this using bounded coordinates. Z starts at zero (the point at which the X and Y axes intersect) and C corresponds to the pixel.

---

[121] Named after the French-American Mathematician, Benoit Mandelbrot (1924-2010) who discovered them. Like Alan Turing he was convinced that nature was based on a set of simple mathematical equations and rules that utilise feedback.

It combines a real number with an imaginary number – multiples of *i* (the square root of -1).

## Fractals

The Mandelbrot diagram is considered to be one of the most important pictures in science and mathematics and has been called the thumbprint of god.

The area of the set is finite – it fits inside a circle with a radius of 2. The length of the border of the image is infinite. This is strange enough, but if you take any part of the border its length will also be infinite.

No part of the border is smooth. If you have high enough resolution and you zoom in, you will find an infinite number of outlines similar to the first outline and if you can zoom into these, you will find the same thing to infinity. Benoit Mandelbrot coined the term ***fractal*** to describe this property.

## Self Similarity

The Mandelbrot set is similar at all scales. It is self similar. It uses feedback to create complexity. It is helping us to understand a fundamental underlying principle of nature. It explains why coastlines, snowflakes and trees have the shapes they do.

## Recursion

The Mandelbrot equation is recursive. It is something that hoists itself up by its own bootstraps. It is a part of itself. It is a function or method that calls itself within its own code. It is something that is part of its own definition.

If you stand between two mirrors you will see an infinite recursion.

## Chaotic Learning

- ✔ Order and chaos are not opposites. They are intimately related. They are different expressions of the same thing. Order and chaos are derived from the same mathematical rules. Complexity springs from simplicity and simplicity springs from complexity. Chaos results in order and order results in chaos. Consider a flock of starlings or a shoal of fish all following simple rules but as a collective exhibiting complex behaviour. If you want to you could consider the body made up of multitudes of cells and complex organs resulting in the idea of a single individual.

- ✔ Things are both simple and complex at the same time. Reality provides different points of view. What appears to be a chaotic random mess of life and shapes in nature is the variety resulting from simple mathematical rules. It can be understood but not predicted. These same rules apply at the level of chemistry[122], coastal erosion and shapes throughout nature and the natural world.

- ✔ Feedback from the environment drives the execution of the rules. Evolution is a prime example of chaos theory in action. Simple rules of replication based on environmental selection have produced the variety of the natural world.

## Management Chaos

Chaos theory says that you can predict the sort of thing that will happen but not the exact details.

A command and control, big design up front process tries to predict exactly what will happen and thereby constrains what can happen.

---

[122] See the Belousov–Zhabotinsky reaction discovered by Boris Belousov dismissed by the scientific community and rediscovered by Anatol Zhabotinsky. It is a chemical reaction that relies on self organisation to produce a pattern of waves. At the time it appeared to break the laws of equilibrium thermodynamics. The chemical reaction doubles back on itself oscillating between states to produce intricate wave designs.

A solution has an evolving shape that is unpredictable because that is the nature of nature. Trying to control this with process is like trying to grow a tree into a straight pipe – you won't get as much useful tree as if you just let it grow as the rules of nature and the environment dictate.

If you build a process strait jacket in advance, you are going to have to force the solution into a shape that it never meant to take in order to fit it into the process. Much of the solution space will be taken up by process matter.

It is by this mechanism that many genius solutions have been, and continue to be, lobotomised at birth.

Organisations that allow for unexpected benefits and solutions that throw new light and new understanding on problems tend to thrive.

### Pixar example

Pixar Animation is one of the most successful companies in the world[123]. Their films are loved by the public and the critics alike. George Lucas describes them as a company that hits a home run every time they bat. How do they achieve their entertaining, intelligent and unique films?

- ✔ They create stress free environments for their staff.

- ✔ They only use technology that allows people to be more creative.

- ✔ They have a culture of collaboration.

- ✔ They only accept ideas from their own staff – i.e. they value their own staff's judgement – this creates both ownership and consistency of vision.

---

[123] "The Pixar Story" directed by Leslie Iwerks in 2007 is a fascinating documentary that gives an interesting insight including interviews with Pixar managers and employees with comments from various industry experts. I caught it on BBC but it is also available in the DVD box set: Pixar Ultimate Collection.

- ✔ They put a high value on patience and skill and they do not cut corners to get things done more quickly or more cheaply if it would compromise the product or the creativity of the artists.

- ✔ They do not separate adults and children in their target audience – i.e. they refuse to be driven by categorisations.

- ✔ Their vision is for people to produce each film as something they will be proud of for the rest of their life. They keep doing it until it is right. In the case of animation it is considered good team work if it looks like it was animated by only one person. They consider it great if it looks like no one animated it.

- ✔ The animators are encouraged to experience what they are going to animate. If they are going to animate racing cars, they go racing.

- ✔ Every idea is developed, nothing arrives fully formed. They encourage feedback. They involve everyone in storyboarding, scripting and voicing. Their consistent message to their staff is *"we trust you"*. They believe that this encourages and leads people to excellence and wonderful things. Trust gives people more responsibility than any carrot or stick could ever do.

In short, change is not predictable. Control obstructs, interferes with and disrupts. Trust, collaboration, patience and skill frees, encourages and amplifies.

### *Straight lines and best routes*

Control processes obstruct, interfere with and disrupt the unconscious skill and conscious effort from which all creative solutions are born. Protected self organisation frees, encourages and amplifies the unconscious skill and conscious effort from which all creative solutions are born.

Management that ignores this scores a straight line in reality and declares it to be the best route between two points. Now imagine a river running down a mountain. Water is about the best thing in the world at finding the most effective and efficient route. It learns the topology and negotiates obstacles to find a route.

This is rarely a straight line. A straight line would involve flowing uphill over obstacles or waiting until the backlog is high enough to reach the level required to flow over all obstacles. It might involve drilling and blasting a straight path thereby delaying the inevitable solution of getting all the water to the bottom. While the drilling and blasting is going on no water flows but we foolishly pride ourselves with being more efficient. More effective water might take a longer route but start supplying earlier.

## *Protected self organisation*

It is important to understand what self organisation means before you inflict it on your teams. It is essential to understand the difference between boundaries and constraints[124]. It is essential to view the trousers of reality and nexus management through the lenses of chaos theory and self organisation.

One of the aspects of chaos we have only referred to obliquely is that of self organisation. Self organisation in chaos theory means that individual parts of a system follow simple rules and feedback from the environment with no controlling designer. The result is natural selection and evolution. The result is unconscious complexity from simplicity. The result in nature is spectacular beauty, functionality and infinite adaptability.

Self organisation is a part of chaos and as such is not predictable. This is the reason why systems such as the financial markets do not find balance or stability through self

---

124 "Boundaries and Constraints" Page 192

organisation, self regulation and self direction. It is a common misunderstanding that self organisation leads to stability. It leads to balance. As this book has gone to great pains to point out, balance is not a state, it is a process.

Self organisation in teams does NOT mean self direction. It means working within a set of defined and essential parameters to achieve a goal. It may involve, and often does involve, uncovering the goal and adjusting the solution in light of real feedback generated by the activity.

It may sound like a paradox, but self organisation requires strong leadership. It requires a person or persons who can create boundaries and get rid of constraints. It almost always requires a leader who can protect the boundaries so that the self organising team can flourish. I call this "protected self organisation".

### *Why self organising?*

Chaos recognises that, for even the most simple mechanistic system, the initial conditions can never be well enough understood for predictability.

The smallest error in understanding the starting point means that, with each turn of the system, reality gets further and further away from what you expect it to be. The rational response is a self correcting system. That is one that is able to determine when it needs adjustment and is capable of making adjustment without losing velocity. The initial requirements are no use if they are the cause of the system going off beam.

Very often processes exist in organisations with the express purpose of making sure that people do not adjust the initial conditions. Contracts and penalties exist to protect the initial conditions. However what they protect is a perception of the initial conditions. Chaos asserts that the initial conditions can never be known well enough.

As many projects progress, the knowledge they uncover is knowledge about the initial conditions. Hindsight is 20/20.

What naturally happens is a process of denominalising the requirements. Requirements are transformed from a list of nouns to a series of verbs. Implementing means denominalising. When this happens the project has moved from what to how and the initial conditions are tested.

## Consequences Of Chaos

The relationship between chaos and order described at the heart of the chaos theory is illuminating. The same mathematical rules produce both. Simplicity and complexity are similarly bound. This is the principle of counterbalance.

In the following chapter I will introduce a principle I call soften to harden. It uses the knowledge that chaos and order coexist and that simplicity and complexity coexist in a complementary symbiosis. If you have one you must have the other. If one side of what you are doing is hard you must soften the other.

This knowledge gives you a tool. If you have complexity/order, then you can be sure that there is a corresponding viewpoint of simplicity/chaos. If you have simplicity/chaos then you can be sure that there is a corresponding viewpoint of complexity/order. The existence of one viewpoint assures the existence of the other.

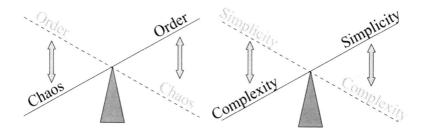

## Complex Problems Usually Have Simple Solutions

If one side of the equation is complexity, the other is simplicity and if one side is chaos, the other is order. The consequence of this is that you handle complexity with simplicity and simplicity with complexity.

In practical terms this means that if you have a complex system to manage and you create a complex process, one side will be forced to simplicity. If you insist on a complex process, the system it produces will not be able to deal with a complex reality – too many false predictions will have been made and it will be substantially flawed by being based on erroneous initial conditions. It will probably be a complex system but complex in all the wrong places. Instead of being solved the problem becomes complexity squared.

If you want to produce a product or system that can deal with the complexity of reality, then you need to produce it with simple processes that do not try to tie progress to estimates of initial conditions. Complex processes produce systems capable only of dealing with simplicity; simple processes produce systems capable of dealing with complexity.

How often do we find that the complex problem we are worrying about has a simple solution? We smack our forehead and whisper "Of course!"

## Command And Control Is A Complex Solution To A Complex Problem That Requires A Simple Solution

Command and control[125] appears to produce chaos in the sense that control is only a surface structure which constrains and quickly loses track of what is really required. It is like an uncomfortable cast iron suit of armour worn by a fat man that limits movement and invites conflict. Trust based, self

---
125 Totalitarian management style described in more detail in Volume one.

organising approaches appear to produce order in the sense that control is applied as an evolving, inherent, attribute – like the instincts of a well trained judo black belt capable of disarming danger and avoiding conflict.

It is a matter of remembering: chaos and order are just different manifestations of the same simple rules that govern the basic principles of reality. Simplicity and complexity are consequences of each other just as order and chaos are the consequences of each other.

## Heuristic And Algorithmic Work

Heuristic work is complex. It is work that requires creativity, expertise, knowledge and problem solving. The more you try to control it with enforced order, the less successful the heuristic work will be. The reason is that the complexity of the work is infinite and a system that that tries to predict all aspects of the work is complexity squared. It can't be done, or even if it could be, the cost would be much more than the work is worth. All that complex processes achieve is to create constraints but those constraints can never take the shape of what you actually need. Knowledge work is heuristic work.

Algorithmic work is simple. It is repetitive and, in order to give it the consistency it requires, it needs firm control and complex process that cover all possible variations to ensure that it does not vary. The reason is that it is possible to create a process complex enough is that the variation is finite.

Even at that, you are still dealing with human beings and the initial conditions can be unknown. This means that around the process for the work itself there must be a flexible and adaptive process for dealing with the people who do the work.

## Processes Can Get Stuck In A Fight Or Flight State

We notice that when there is great danger, such as war, heuristic behaviour is discouraged and algorithmic behaviour is enforced. Growth comes from creativity, breaking out of the stimulus-reaction[126] cycle and introducing learning and progress. We are willing to sacrifice that when dealing with emergency and real and present danger. It is a type of systemic or process fight or flight response.

In the biological fight or flight response all the blood moves to the muscles away from the internal organs and the thought process until the danger is over. Organisations react the same way when they are in danger.

Stress and often illness is the result of being constantly in the fight or flight state. It is an emergency reaction that should only be called as a last resort. The more quickly we can start thinking in a dangerous situation the better. The same applies to organisations. Fight or flight processes and management styles divert energy away from the creativity and adaptability required to negotiate the chaos of the marketplace.

## NLP As A Chaos Utilisation Tool

NLP provides some valuable tools for dealing with the reality of chaos. It always assumes the simple/complex relationship. Simple problems can have complex causes and complex problems can have simple solutions.

One way it differs from other approaches is that it does not try to provide a detailed solution. Instead it relies on the way the creative unconscious mind solves problems. It allows it to self organise within careful boundaries. It agrees boundaries,

---

126 Stimulus-reaction cycle – I have a headache I take a pain killer. I have a headache-I take a pain killer. Sooner or later you need to break out of this and find the cause of the pain rather than just dealing with the symptom. This breaks the stimulus reaction cycle and introduces learning and probably change.

removes constraints and leaves the complexity to you. It looks for the simple view and manipulates that, with the knowledge that at the other end a lot of complexity is being moved about. It trusts the organisation or brain to find the solution. It knows the kind of thing the solution is, but not the exact thing. In this it is consistent with chaos theory.

A good coach does the same thing. A good coach will not arrive with a detailed list of behaviours for you. A good coach will arrive with a set of principles and trust you to apply them and help you listen to the feedback and experience from yourself or your organisation.

## Attributes Of Reality Viewed Through Chaos

- ✔ It is recursive.
- ✔ It is self similar at all levels of scale.
- ✔ It is self organising.
- ✔ The same underlying principles that produce unpredictable behaviour also create the patterns we recognise as order.
- ✔ Simplicity and complexity are counterbalancing consequences.
- ✔ We don't really know how or why it works but it does.

# Thinking Inside The Box – The Treehouse Of Truth

*"Don't lose your confidence if you slip,*
*Be grateful for a pleasant trip,*
*And pick yourself up,*
*Dust yourself off,*
*Start all over again."*
**Dorothy Fields**

## Boundaries, Counterbalance And Change

Whether you swing a double headed axe left or right it cuts. It can be used to chop firewood or chop off your foot. Consider bars on a window. The bars on the window can be used to imprison you or to keep danger out.

This kind of duality exists in nature and in reality. *The Trousers of Reality* is about duality in the sense that logic depends on paradox. It depends on the flow between order and chaos. It depends on the flow between simplicity and complexity. There are points where they are balanced but they are temporary and they are moving.

We can find it difficult to admit that all human endeavour is balanced on the knife edge of entropy and chaos. We pretend we can predict and control. Prediction is always a guess and control is always an illusion.

Everything is double edged. This effectively places all of us right at the centre of a paradox. It can be difficult to take a stand when you know that whatever you are standing on can suddenly become the ceiling rather than the floor. So how do we survive and function?

## Creating Space For Stability

We cannot see the edges of reality, so we build virtual boxes and pretend they are the edges. This is a great idea. It is among our finest abilities.

Because infinity, chaos and time are all a bit too big to handle, our mind, quite sensibly, puts up temporary walls, floor and ceiling so that we can get things done.

We are like children in a tree-house. We know that the ground is not really up here in the tree and that the walls of our hut are there to keep us from falling out. We have put them all there so that we can create a set of boundaries and play our game within them. At any time we can scamper down the ladder to the ground and face growing up.

As we grow up we learn to create other boxes that allow us to deal with sets of problems. They help us to avoid becoming paralysed by the enormity of reality. We construct the tree-house of truth.

## Willing Belief Is A Tool

This willing suspension of disbelief is a skill. It is a talent bequeathed to us by evolution.

Could you watch a film or a play without allowing yourself to believe temporarily that it is all true? We live a good book, a good film or a good play. The attraction is the way hum drum reality and unimaginative rules can be replaced temporarily with the story and the alternate rules. These rules are consistent often only in a make believe world. We can grey out what we don't want to see and live in that brightly lit box for the duration. We know that when we wake up from the dream that we will be able to let go of these strands of willing belief.

This is the same tool we use to deal with our professional and private lives and to deal with our mortality. If it were not for this we would be paralysed by our own consciousness.

## Make Sure The Tree-House Has A Ladder: Reinstating Disbelief

As with all tools or skills that are worth having, this one is double edged. Sometimes we can fool ourselves so completely, suspend our disbelief so thoroughly, that we forget that the boxes are constructs.

Sometimes we recognise these conditions as phobias, beliefs, superstitions, processes and other constraints.

It is a skill to create these fictions and to suspend disbelief until the fictions are no longer needed. We rely on our ability to create boundaries. It is difficult to even name a thing without defining its boundaries.

As long as we remember they are constructs, we are chopping firewood. As soon as we forget they are under our control, we start chopping off our feet.

The boxes are recursive functions. The box uses itself until it is no longer needed.

A test would be useful, so that we know when the box is no longer needed and that the danger has passed. We want to be left with increased skill and knowledge when the initial conditions that led us to create the box have moved on.

So, we need to doubt the existence of boxes and we also want to recognise them.

We all have limiting beliefs as a result of suspending disbelief, probably initially suspended for a good reason. We can restore choice of behaviour by challenging these limiting beliefs and reinstating disbelief. (E.g. the disbelief in limitations on our inability to learn.) There are other limiting beliefs that are a result of suspending belief in our abilities. Once more we can reinstate belief by suspending disbelief and allowing the possibility. (E.g. in our ability to implement some desired change).

## These Beliefs And Behaviours Are Strategies

Strategies can be reprogrammed. In other words:

- ✔ Doubt the boxes that restrict.
- ✔ Upgrade the boxes that are useful.

At a very basic level NLP teaches us to use a set of techniques to do this. By following the easy steps you can learn to deal with phobias, unproductive behaviours and a host of common fears and anxieties.

Both strategies and limiting beliefs are boxes. Strategies are boxes that are still useful and limiting beliefs are boxes that have become a liability.

## Learning Is A Recursive Process

Here comes the inevitable recursion. In order to learn anything, even something like NLP that challenges limiting beliefs, we have to construct these temporary boxes. NLP calls them presuppositions. The trick is to recognise them for what they are, as part of the process, and to let them go when you have learned enough to move on.

Part of learning is to identify where, how big and for how long we need boxes. We develop judgement about which ones really need demolishing if they are starting to hinder fact based thinking. This is the art that evolves from practice.

At the first stage of learning anything it is possible to confuse the glimmer of inspiration with the resourcefulness of the craft and to be over impressed by the initial rush of new ability – or beginners luck. It is important to remember that the canvas is supported by a frame. Inspiration, skill and craft are interdependent.

## Phobias Metaphor To Highlight The Importance Of 360 Degree Learning

If an NLP practitioner were helping somebody with a phobia of heights, they would need to concentrate on more than the exciting technique that might remove the fear. They need to make sure that as the phobia is removed, the person will remain safe climbing ladders. Nobody wants to be so unconcerned by heights that they forget all about safety. The technique needs to be tempered with thought about the ecology of the situation and the consequences of the technique working.

If I never climb anything higher than a chair, I need one sort of box about heights. If I am a tree surgeon, I need another.

The same goes for most phobias, be they fear of spiders, clowns, snakes or long words[127].

The phobia is a box that was ill constructed. It can be described as an inappropriate response to a stimulus. There are appropriate responses and removing the response completely can have disastrous consequences.

If a person is a trained arachnologist, they know which spiders are dangerous and which are safe and under what conditions. The box they need has very transparent walls.

Most of us would not know a venomous spider and therefore when we are in countries where there are dangerous spiders, we need a much stronger box to protect us. We certainly need to treat all spiders there with respect and not trifle with them.

Then there are conditions when you do not need any box at all. Perhaps you live in a country where there are no

---

127 Sesquipedalophobia - so you can't even tell them what is wrong with them :)

venomous spiders at all. They are benign fly catchers and pretty useful allies.

It is the level of knowledge and the context that dictates the strength of the box required.

# The Principle Of Counterbalance

## *Soften to harden*

I discovered this metaphor through learning Tai Chi and Skiing. As one boundary hardens, its twin softens.

> When you are learning how to ski you move from the snow plough position[128] to parallel skiing[129]. As you do this, you realise that whatever it looks like, it feels totally different. It feels like an unexpected cycling motion in your legs as you transfer weight smoothly from one ski to the other.
>
> In order to turn you put your weight on the ski that is going to describe the outside of the turn. As you do this you take your weight off the other ski. Inevitably you approach the next turn and you prepare the inside ski to become the outside ski. You are constantly softening and hardening in a smooth and gradual transfer of weight from one ski to the other. This softening and hardening is what provides the control. It allows you to control your angle to the hill, the size of your turn and your velocity. As you advance, your tracks turn from angled Z shapes to smooth round S shapes.

---

128 A skiing position for beginners. You angle the skis in a V position with the point aiming downhill. The angle of the V, which resembles the blade of the eponymous snow-plough, determines your speed. This gives you the simplest way to control your speed so that you have a method to ski without having to have much skill. Doing this builds confidence and prepares the way for more advanced techniques.

129 A skiing position in which you hold the skis parallel to each other and control your speed by your angle to the incline of the hill and traverse by moving from side to side across the piste at your chosen velocity with much more flexibility of movement.

Another thing that surprises beginners is that you control the speed at the end of the previous turn and not the beginning of the current one. How you end the last turn determines how you execute the next one[130]. Your skis are forethought and hindsight.

Beginners try to put their weight on the inside ski, the one closest to the hill, and fall over a lot. In order to progress, you must have the confidence to lean away from the hill and trust your skis to do the job they were designed to do. This feels counter-intuitive to the point of madness to a beginner. You have to unlearn what you think you know about balance. The rules of gravity still apply but the context is very different. You need to learn different reactions to the same stimulus because the old ones will trip you up.

When it comes to methodologies people tend to have the same reaction. As they enter a new context, they want to lean on old reflexes. They blame the new methodology for the tumble. The instructor may have just told them that they must trust the principles and put their weight down hill. A skier who ignores this advice and leans back toward the apparent safety of the hill will find his or her skis shooting ahead and his or her body falling backward as Newton's third law of motion kicks in.

For many management methodologies, this lunge toward security is a lunge toward control and authority. It seems natural to keep close control over people and resources. It seems a safer route to control costs and insist on restrictions and forbidding everything that is not explicitly allowed. It seems more managerial to throttle back the resources in order to make people compete and throw up winners.

This behaviour exacerbates the problems it is trying to address and the new approach is often blamed

---

130 This has a curious resonance with the idea of retrospectives in project iterations.

### Control and creativity

When you have hard control over people their creativity softens and becomes ineffectual. If you want hard creativity, you must soften control. Restrictive processes may give you a sense that no one is doing anything you don't want them to do, but it also ensures that no one will excel. They are all too busy gaming the system or struggling to survive in it. Trust people to do the job they were trained to do and you will execute smooth effortless progress with controlled velocity.

### Testing and maintainability

Even at the level of programming computer code this applies. If your code has no tests or tests that are too far removed from the function of the code, the code becomes restrictive, monolithic and hard to maintain. It has to try to second guess every eventuality and deal with all error conditions that it might be tested for or encounter while running. This makes for complex and inefficient code.

Code that evolves within an environment of comprehensive unit tests based on positive requirements tends to be agile, efficient, easy to maintain and better designed for purpose.

### Not all trade off is negotiable

### (especially if you want to have your cake and eat it)

Many organisations want both. They want creativity and they want to control every action of their employees. Many software development departments want generic test tools late in the process, but also want easy-to-maintain code.

If you try to explain this as a trade-off, people immediately begin to negotiate for position. Once people establish a position they tend to defend it. They pull the lid of the box closed and turn work into a childish king of the hill game.

Any management methodology needs to be fluid. There are no positions to defend. There is a flow of context that needs to be understood. The reality of any situation requires you to be able to harden and soften the complimentary and opposing muscles[131] of the organisation in an ongoing process of balance.

## *Using feedback and filtering*

The mechanisms for balancing any given situation or project are feedback and filtering. They are different functions of the same tool.

Whether we talk about control, power, love, wealth, identity or reality, we rely on feedback and we are always filtering the sort of feedback we find useful.

From our choices we create boxes, models or constraints as we seek to control fear, grief and anger in one context; and scale, time and difficulty in another.

What if we make the wrong choices and filter out all the relevant feedback?

## *Unnecessary prosthetics create dependencies*

A cast on a broken leg makes mobility possible and encourages the bone to heal properly. It also causes the muscles to weaken as the bone heals. I know because I broke my leg when I was a child and inside the cast the leg muscles shrivelled as the cast immobilised them and the bone healed. When it came time to take the cast off I was horrified at my

---

131 Complimentary muscles that are involved in the same movements and enhance the efforts of the other e.g. back and biceps in a pulling exercise. Opposing muscles are muscles that perform opposite actions. E.g. biceps and triceps. Biceps pull the arm in and triceps straighten it. They pull the arm in opposite directions. While one is working the other is resting. While one is hard the other is softening. If they were both to contract at the same time there would be very restricted movement only in the direction of the strongest muscle.

pale skinny leg. I could not stand on it and I doubted that I would ever walk again. I wanted the cast back.

Of course by that evening I could stand on the leg and within a few weeks it was back to its old self carrying me up and down ladders to tree houses.

Imagine a hospital where only some feedback was taken into account and people refused to learn anything about the healing process, muscle regeneration or bone knitting beyond surface level symptoms and initial conditions. In this sort of hospital when the cast is taken off after six weeks, the leg appears to be damaged beyond repair based on observation. Back on goes the cast and after a couple of years of this the muscles have completely atrophied and leg really is beyond repair.

> The point of this metaphor? The cast is process; the bone is repeatability, predictability and control; and the muscle is creativity, ingenuity and judgement.

The patient continues to be delighted by the cast because without it the leg can no longer support any weight. There are other consequences if the cast is left on long enough and the patient acclimatises to walking with it and subsequently finds it difficult to imagine life without it.

> Organisations that are dependent on process and technique rarely recognise the limitations those things are inflicting and they are very fearful of having them removed. They have forgotten that people will take up the slack.

The cast, once necessary, is now stopping any running, swimming or climbing of trees. There are people for whom this sort of prosthetic is necessary but human beings are hugely adaptable so they get on quite well with their lives. They would also be the first people to say that if you don't need a prosthetic, don't use one. They also welcome the invention of new, lighter and more flexible prosthetics.

### Adapt behaviour and strengthen skill

Looking into the face of reality can be a frightening thing. It is useful to be able to create avatars and straw men. We can contemplate that there is more than one infinity. We can grasp the idea that there is an infinity of infinities and that even this set is only an infinitesimally small subset of the set of infinities that exist. Trying to look into that abyss drove mathematicians like Gödel to the edge of sanity and the edge of health.

> Many organisations have to deal with their own overgrown size, the complexity of business reality or their own insecurity and feel compelled to create an alternate and more manageable reality in which to suspend their people. Often these are based on rules devised for some previous incarnation of themselves and have grown like Gordon Mackenzie's giant hairball. Many people take these alternate realities way too seriously and inflict very real consequences on the real people who work there.

The box is a safety device but it is not reality. Reality is so stupefyingly huge that our very identities are boxes to help us cope. It is fine to create short term boxes which are fit for purpose, but once their usefulness is over we need to disassemble them altogether or to reshape them in line with the new requirements.

> Process is one of those boxes. We apply process when we do not understand what we are doing. We look to process for guidance. Corporate culture and processes need to be constantly challenged and connected to their consequences.

### Spiders metaphor for knowledge contexts

Think about our spiders again. If we know that there are some venomous spiders but we do not know which ones they are, it is rational to assume all spiders are venomous even though you know it to be untrue. Until you know for certain it is better to treat them all as guilty until proven innocent.

However, and it is a big, big however, you need to find out which ones are venomous and which ones are safe or you are making your life unnecessarily stressful as there are spiders everywhere.

You need to gain skill at observing them and information as to how to recognise the dangerous ones.

## Process And Feedback

Any process that has outstayed its usefulness is quite similar to a phobia. To be more accurate, the pathological need for process is the symptom of a phobia. A good candidate for the actual phobia is Metathesiophobia or perhaps Sophophobia or maybe Tropophobia or even Kainophobia or likely Kakorrhaphiophobia. (Fear of changes, learning, making changes, novelty, failure). Maybe it is just plain old Panaphobia (Fear of everything).

What exactly do we think we are protecting ourselves from with these processes? They are nothing more than corporate prosthetics.

Processes are designed for approximations. When they no longer deal with the situation at hand, they are much more limiting than a cast on a healthy leg.

Use processes until you learn, but remember that you learn through doing. If the process stops you doing things, then you must consider questioning the process.

Here are some rules about the boxes often used in business:

### *Rules about boxes (corporate prosthetics)*

Meta rule: business prosthetics should be working themselves out of a job.

- ✔ A really good process should be teaching people to work without the need for a process.

- ✓ A really good coach should be teaching people to listen to their own feedback and create a learning environment by becoming each others coach.

- ✓ A really good manager should be managing people into a situation where they do not need managing.

- ✓ A really good technique should be teaching people to integrate an effective skill as a habit.

We know from neuroscience that tiresome and boring activities cause the brain to succumb to the plastic paradox and atrophy, while new and challenging activities change the structure of the brain. All learning involves changing the structure of the brain. The prosthetic should encourage permanent learning but many processes do exactly the opposite.

### Bach Y Rita's balance machine

Paul Bach Y Rita dealt with a case of a woman with a damaged vestibular system[132] – the part of the brain that deals with balance had been damaged. She had no sense of balance and was always falling. It had been declared an incurable condition.

Bach Y Rita created a vestibular prosthetic apparatus that created tactile feedback on the tongue that indicated her angle with relation to gravity. With this prosthetic she had perfect balance.

They found that if they took it away after one minute she got 20 minutes of residual balance without natural or artificial sensors. 2 minutes use gave 40 minutes of residual balance when it was taken away. 40 minutes of using the apparatus gave her 3 hours and 20 minutes of balance without it. As her brain got the message and found secondary pathways to

---

132 Detailed in "The Brain That Changes Itself" by Norman Doidge.

replace the damaged ones eventually she did not need the apparatus at all and regained her own balance.

### *Process design*

This begins to explain that doing some things certain ways gives independence from process and constraint. It indicates how we should be designing process. Many processes are like grass ticks, a kind of parasite that burrows in deeper and deeper to suck the blood of the host. We should be designing process to become redundant. Even the tick gets full and falls off.

Processes should be designed to generate and provide feedback that is useful to the host. This means identifying pruning away feedback that is useful only to the parasite process. Who cares how healthy the parasite is: how is the host doing?

We are feedback machines. We are evolved to process feedback. Anything that corrupts that feedback is dangerous. Anything that focuses that feedback and filters it is helpful.

There are two boundaries we need to explore.

✘ Not enough feedback?

✘ Too much feedback?

## Feedback Engineering

### *Not enough feedback*

People with nerve damaging diseases like leprosy or diabetes are totally dependent on process. If they have suffered neuropathies they may feel no sensation[133] in parts of their body. This is often the feet. They are likely to damage themselves unwittingly. Accordingly they follow a process of

---

133 Pain for instance is a useful feedback that warns us of impending damage.

scanning their feet regularly and following processes for even simple things like checking new shoes for sharp edges and not walking around in bare feet. They create useful boxes with boundaries to manage what the rest of us just rely on the feedback from the nerves in our feet to tell us.

When we are not getting enough feedback we simulate feedback. Like the person with numb feet using their sight and their hands to bridge the gap, we learn to look for an alternative feedback and to glue it together with process. This is never quite as effective as the natural feedback. You can be really good at watching out for sharp things but it is never going to be quite as good as having the feeling in your feet and hands.

## The problem with simulated feedback

The way some processes try to simulate feedback from an organisation is like trying induce running ability in an athlete by driving him around a running track on a motor bike to experience speed. Partaking in enforced and artificial feedback for a process has about as much effect on people's abilities as being driven on a motorbike will have on an athlete's ability to run fast. On the day of the race the athlete who has spent all the training time riding a motorbike will find that the required muscles have become sluggish and slow from inactivity.

People need feedback whatever they are doing and if you clog up the feedback pathways with meaningless garbage, the garbage in garbage out rule kicks in via the neural realities of people.

## Encourage real feedback

In order to run faster an athlete has to practice running. Visualisation is a good thing so one trip around the track at artificial speed may offer some remote possibly of help but the best feedback comes from actually running. The only way to learn to run faster is to train the muscles by running and accurately visualising running.

People learn to be responsible by being given trust. People learn to be good at their job by being confident enough to try new ways of doing it.

## *Too much feedback*

If the process is inhibiting the feedback or generating noisome feedback, we are in grave danger of missing valuable opportunities. Most of us deal with information overload (and I discuss this throughout the book). If we feel we are getting too much feedback, most of it is probably relevant only to the parasite processes we have surrounded ourselves with.

## *Make distinctions*

Learning to make distinctions and create appropriate filters is like everything else: we have to train ourselves by doing it. Even at work it can be necessary to apply an appropriate filter that lets through only what you need to know or what is useful. There is often much important sounding feedback that seems relevant but is really just misleading nonsense.

> Example: How the company shares are doing on the market is irrelevant for most levels of management in a company. Responding to the feedback generated from the goldfish-like attention span of the stock market is no way to run a business. The market is all about perception and it is divorced from concrete cause and effect. Because of this the feedback you get from the market about your strategies is not relevant because your strategies do not affect it. The market fluctuates when certain people sneeze and you can't legislate for it. Therefore the feedback would appear to be irrelevant.

Can you tell parasite feedback from host feedback? Is the feedback benefiting the process or the object of the process?

- ✔ Learn to ask yourself: "How does this feedback help me or advance what I am being paid to do?"
- ✔ Learn to adjust your filter as the feedback changes.

## Scaling Process

The more rigid a process, the more difficult it is to scale. Most bureaucratic process is cumbersome and makes assumptions about the domain in which it is used.

Often work-flow and time-management processes assume that people will be able to switch between tasks and not lose any time readjusting. They are quite happy to allocate fractions of people per day or even fractions of people for fractions of hours. This is utter nonsense but people find they have to pretend that it is possible and waste time and effort trying to adjust to the way they are forced to use the system. There is always error when the parasite thinks it is the host. The bigger the domain, the bigger the error. The process will protect itself and protect the error. It will insist that the process is always right and that there is no need to correct it. Often, no way of correcting it is built into the system. It actually uses the flexibility of people to hide its own foolishness[134].

When this occurs in a planning process you can imagine how assumptions, leading to errors and rigidity, can unseat a large project without anybody really being at fault.

The less rigid a process, the more the human brain can take up the slack. Neuroscience indicates that the human brain is more flexible than the most powerful computer. It can use the computer for the algorithmic work of remembering facts and

---

134 Gödel's incompleteness theorems, discussed in Volume one: Working Life, conclude that a system cannot use itself to prove itself true. See also "Mathematics, logic and philosophy" on page 103 of the current volume.

adding up really quickly; but the human brain can do something better than any process or computer. It can bend the rules.

## Bending the rules to apply common sense

The way most things get done is by bending the rules. The trick is to know what rules to bend and when.

In a process the rules are there to protect you from deviation. They can also isolate you from innovation and invention.

This is particularly true in technology and software development, which is all about innovation and invention. The wave front is always ahead of the process. The technology is always progressing and process tends to lag behind.

Versatile methodologies apply minimal process in order to allow errors to be corrected. It is made up of rules which are made to be bent. They will bend a long way before they break.

These approaches can be hijacked by process driven people who often attempt to establish them as a set of rules.

These people treat the rule book as if it were an instruction manual. Progress and performance are often measured by how well the rules are followed.

Versatile processes are not rules, they are suggestions. They are flexible enough to allow opportunities to be taken. As a process they have enough stiffness to hold the shape but not to constrain. They can deal with varying initial conditions and they can roll with the punches.

Regardless of how they are used and appropriated by short sighted process engineers, the principles remain true. You can still use the principles behind Agile, Lean, NLP, TOC etc. Sadly, you may have to call it something else if it has already been implemented as a set of constraints and rules.

Volume three of this series looks at the specifics of how to apply the principles discussed in this book.

## Productivity And The Rights Of The Individual

> In 1936 Charlie Chaplin made a film called "Modern Times". In it his little tramp is working on an assembly line and has to work at an ever increasing rate to turn knobs on widgets as they pass him on a conveyor belt. Eventually he has breakdown and goes wild. He starts tightening everything including his co workers' nipples and the buttons on the secretary's dress. The film is a masterpiece and highlights the dilemma of human beings remaining human and having lives in the midst of an increasingly productivity driven society.

### *Algorithmic work*

Let us just imagine somebody employed to do an algorithmic job. The job is bad enough with constant calls for increased productivity and the monotony of the work. However, they are being paid to do it and they are getting on with it. Now somebody introduces stretch goals and they find that they have to be innovative without the slack to be innovative in.

Most of this book has been aimed at people who do heuristic work (i.e. work that by its nature, requires people to be inventive and creative) and for whom the process is a constraint. In every organisation there are also people who do essential algorithmic work (i.e. work that involves following instructions to the letter). Some people are suited to this for a variety of reasons. They may be recovering from burnout, be temperamentally suited, have heavy demands on creativity elsewhere etc.

Realistically most human activities contain varying amounts of algorithmic and heuristic content. This conforms to my idea of soften to harden. Algorithmic aspects of heavily heuristic content provide a counterbalance and if approached positively can be satisfying and provide insight, feedback and balance. The same can be said of primarily algorithmic activities that contain heuristic elements, sometimes it can lead to thinking about how to build a better conveyor belt or automate the washing up.

## Difficulty Versus Complexity

Difficulty and complexity are often interchanged as if they were the same thing. Very complex things can be very easy and very simple things can be very difficult. Shovelling is pretty simple, but it is difficult if you have to shovel a ton or two of dry sand into a wheelbarrow and move it half a mile down the road. In my first summer job I had to file a five year backlog of licence applications. It was simple but it was certainly not easy – while it was not complex it was very difficult. Reading is a phenomenally complex activity when you consider what is really going on, yet for most of us we consider it an easy thing to do – while it is complex it is not difficult.

Managing well means appreciating what people are doing from their perspective.

## Ambition is a bad measure

Even among knowledge workers there are people who just want to do their job well and do not aspire to promotion. These are people who have applied, passed the interview in good faith and work for the company doing the job that they were employed to do. It is often a job well within their capabilities.

They are happy to be given work to do with management guidance and a clear set of requirements. They will do it and do it well. They do what they are asked to do. They come in on time and occasionally work late. They have lives and interests outside work and do not care about stellar careers. They do not look to their work as the only source of meaning in their life. This is okay and a perfectly valid model of working life in perspective.

This model does not compute with the models of some ambitious managers. They interpret this behaviour as a lack of ambition. Or worse, because they do not see the stress they have associated with effort and productivity in their own model, they can identify this sort of person as lazy and an untapped resource that can be further exploited. Where a more enlightened manager would recognise productive slack, they perceive waste to be addressed.

As a result these balanced workers are often given arbitrary targets and put under immense pressure. The productive work they were doing without fuss and quite often with unnoticed skill, talent and creativity, gives way to the demands of the performance measuring process. This process almost always measures capabilities imagined relevant by people who do not understand the job being measured. The fact that this kind of performance measurement is introduced is tautological proof that it has been introduced by a management philistine.

### *Performance management damages people and productivity*

As performance measurement increases and real productivity dwindles, a truly horrible destructive feedback loop can begin. When the delayed feedback of the diminishing productivity finally registers somewhere, it is not associated with the destructive effect of the performance management. It is most likely attributed to the lack of high scores on the arbitrary capabilities and metrics being measured and so the performance measurement kicks up another gear. Where there were productive, competent people, there are now stressed people pretending ambition and quite often pushed out of their own spheres of expertise into the performance measurement school of management where they can help keep the whole artifice lurching along.

## Corporate intrusion is just opportunism

If a person is hired to do a job, both parties have entered into a contract. Should it really be part of the deal that this person should seek promotion and stretch themselves according to some corporate ideal, especially if they are good at what they do, and what they do is valuable? In large companies, where savage politics are played, most sane people just want to keep their head down and get their jobs done. There seems to be a fashion these days to require people to push themselves to the limit, mentally, physically and emotionally. They are submitted to the most intrusive, subjective analysis and asked to become someone they are not to satisfy some arbitrary target. The deal with the employer was that they were hired for their work and their skills; which they are happy to keep up to date by learning the latest techniques and following developments in their chosen profession. Who they choose to be as a person is entirely their own business and they have a right to their private life. They should never be asked to conform to some middle manager's idea of what their self improvement or levels of ambition should be. Their private time is their own to do with as they please.

## Remember there is more to life than the job

It strikes me that some companies force people to identify with their jobs to an unhealthy extent. This manifests itself in stretch goals and annual, quarterly and even monthly reviews. Some companies ignore the fact that people have competently and loyally carried out their jobs but focus on how they have failed to "grow" sufficiently. This is usually at the whim of some harebrained junior or middle manager trying to show that he had the right stuff by spilling innocent blood.

Stretch goals are applied no matter how much you do. It is assumed you can always be squeezed another bit. If you are not on the edge of a nervous breakdown, you are not giving enough to the company.

Given what we now know about neuroscience and behaviour, we can assert that this is unproductive from the point of view of the individual and of the company. It is very short sighted and ultimately both people and productivity suffer. This type of corporate behaviour is based on a child's game called King of the castle.

## King of the castle

*"Eliminate slogans, exhortations, and targets for the work force asking for zero defects and new levels of productivity. Such exhortations only create adversarial relationships, as the bulk of the causes of low quality and low productivity belong to the system and thus lie beyond the power of the work force."*

**W Edwards Deming**

Companies often have a carefully worded mission statement or vision statement[135]. It is likely to be something like:

"We aim to be the market leader."

Senior managers clap themselves on the back and figure they have earned their salary for the next ten years. They do not want to be troubled by detail, just make it so.

Middle management decides that the best way to be the best is to be the most profitable. They reason that the most efficient way to be profitable is to streamline. In their mind the mission statement becomes "Streamline so that we can compete". What follows this statement is a push to cut costs and, at the same time, increase productivity from the workforce.

This is often, euphemistically, called 'focussing on core business'. This is code for putting an end to anything which is not strictly controlled by the process. Unfortunately this can often include something called thinking.

---

[135] These are usually so well intentioned and so badly executed that whenever I hear about meetings where hours are spent crafting and arguing about them, I literally want to weep and go live in the woods. Somebody has grasped the echo of a technique without understanding the first thing about the principles.

Most of the productive workforce in a modern company deals with knowledge. Managing that knowledge requires thinking. It requires a certain sort of mind to deal with the volume and scope of knowledge. This kind of mind does not react well to being told what not to do.

There is also the small matter that "being the best" is an outlandish target. It is a childish goal. It is equivalent to jumping up on a little hill and shouting "I'm the king of the castle and you're a dirty rascal!" It means nothing. How do you measure it and if you can, and if you are not the best by any measure, has the whole venture been a waste of time?

How about aiming at being very good at what we do while enjoying what we do and enriching the lives of everyone involved in this endeavour. I like Pixar's vision statement to produce something of which we will be proud for the rest of our lives.

## Throwing An Exception

I know. I know. There is always the other game to consider. Book four in this series grasps that nettle. For now let's deal with the main logic branch and throw an exception[136].

---

136 In certain styles of computer programming the main body of the code is written as if everything will always work. The code tries to work and if it works it moves on. If there is a problem it throws an exception. This is a message to the error subsystem with details of the error and where it happened. It does not try to deal with the error itself. This allows the error handling to be decoupled from the main body of the code. This allows the programmer to concentrate on the main purpose of the code. It makes the essential logic of the program clear. It also allows the possible error conditions to be handled separately and not become entwined and confused with the "happy path" - as the part that does not deal with errors is called. It means that error handling can be updated without affecting the core logic.

# Metaphor – Levels Of Modelling

*"There are few things as toxic as a bad metaphor.*
*You can't think without metaphors."*
**Mary Catherine Bateson**

## Flow

The concept of flow has become pretty universally acknowledged by programmers, musicians and sportspeople. It is that state where we can find ourselves "in the zone". We are firing with all thrusters, one with what we are doing and digging deep into our resources and capabilities. It is almost as if the barriers and inhibitions of daily life drop away and we can access the pulsing jewel at our very core like some otherworldly drive.

What if you could go into flow at will? What if you could influence the people around you to reach that state?

Can you remember a time when you were totally in flow? What was it like? What were you seeing? What were you hearing? What were you feeling? What happened just before you went into flow?

Just thinking about those questions could be making you recall and reach out for that state right now.

At the heart of any investigation into reality we find flow. It may be called superconsciousness, transcendence, higher state, human potential and a host of other labels. Some of these labels create an adverse reaction and some of them are helpful.

## Labels

We have attached values and beliefs to the labels. While values and beliefs enable us to function they can also be limitations. They define the metaphor and obscure the thing itself.

A label is a thing's surface structure. We use language, our most amazing tool, so that we can literally put handles on concepts and move them around. The clue is quite often under our nose and recursive: What's your handle? How can we get a handle on that? Can you handle it?

William Shakespeare is not called the greatest playwright without justification as his insight into the human condition frequently proves.

> "'Tis but thy name that is my enemy;
> Thou art thyself, though not a Montague.
> What's Montague? it is nor hand, nor foot,
> Nor arm, nor face, nor any other part
> Belonging to a man. O, be some other name!
> **What's in a name? that which we call a rose**
> **By any other name would smell as sweet;**
> So Romeo would, were he not Romeo call'd,
> Retain that dear perfection which he owes
> Without that title. Romeo, doff thy name,
> And for that name which is no part of thee
> Take all myself." [137]

## Metaphor And Epistemology

What's in a name?

A name is a metaphor. It highlights the mental processes of language. Much has been written about how language influences our thinking. This has had a huge influence on the study of epistemology – the study of knowledge.

---

[137] Romeo and Juliet Act 2 Scene 2

Metaphors and surface structures are pitons in the walls of reality. They allow us to shrink reality down to the size of a chessboard and move things around.

We gradually bind together the metaphor and our understanding of the real thing, in a tighter and tighter loop. We ignore all evidence that does not support this metaphor. We have taken the most useful tool ever created to deal with reality and trapped ourselves inside it. Immediately we have to invent deus ex machina so that we can keep moving. The recursive cycle pulls us down the vortex.

## Values And Beliefs

Our values and beliefs are metaphors for ourselves. We name ourselves and ignore the wisdom of Shakespeare. We limit ourselves two levels down. Count them.

So we have created metaphors for ourselves and everything else in the universe. We literally create the universe in our mind, or at least a reverse avatar for it.

This is existence and quite a brilliant way to deal with the immensity of reality. As long as you remember to take the red pill[138] from time to time.

## Problem Metaphors

What we term problems are those times and circumstances when the metaphor, or surface structure, and its properties can no longer contain the reality, or deep structure of something. This can be anything, your job, your marriage your self. The name, the label, the metaphor no longer defines what the thing itself has become or what you know about it. "Montague" no longer defines Romeo. There is a tension[139].

---

138 Ref to Matrix metaphor that life offers us choice : the red pill is truth and the chance to perceive reality at another level and the blue pill is wilful ignorance.
139 This is an imbalance and as we discovered in volume one, just as nature abhors vacuum it also abhors imbalance.

One route is to seek to make the metaphor stronger. Psychotherapy, traditional management techniques and some interpretations of science[140] take this route. It only delays the inevitable and makes the reckoning more severe.

A more creative use of the tool is to realise that it is a case of a metaphor that has outgrown its usefulness and needs to be updated or replaced.

## Changing The Metaphor Uncovers The deep Structure

In this changeover between metaphors we can glimpse the thing itself for a moment. Transcending is not being above it, it is being part of it. As we go through the process of letting go of the metaphor and reach for the next one that will allow us to contain the reality for a while, we have to transcend.

We merge with the thing because for an instant it has no protective wrapping. Many, if not most, people have had moments like these. They have experienced moments of pure inspiration. They have experienced moments of pure clarity. They have experienced moments of pure transcendence. They have glimpsed their own true nature beyond the metaphors and beliefs they normally maintain.

## Leaping Between Metaphors To Access Flow

Colin Wilson talks about his friendship with Abraham Maslow and tells how Maslow became very interested in these peak experiences. He gathered his students and started to talk about them. As the students talked about and remembered peak experiences they began to have peak experiences. As the study went on they began to report more and more occurrences of them. Maslow concluded that these altered

---

140 When new ideas from Newton to Bach Y Rita try to enter science, even by its own rules, but threaten to destabilise comfortable theories with hard evidence, inertia lies on the side of the established metaphor.

states should be studied and cultivated. He said that ""Peak experiences are transient moments of self-actualization." [141]

Peak experience and flow is something we can learn to access at will. It can be anchored and triggered[142].

When you are in this state time drops away. Awareness of bodily sensation drops away. You no longer see the problem you are working on. You become one with the problem. You inhabit it totally. Only the flow of the thoughts that bind the solution exist and you flow into them releasing, unblocking and routing as you go.

The metaphors retreat and you are left with the thing itself. For a while you see the labels as labels and can look at what they point to.

It does not require belief. In a way it is the opposite. It requires a simultaneous suspension of belief and disbelief.

When we leap from metaphor to metaphor and flow into the space in between, we have an opportunity to triangulate. This gives us perspective.

## Metaphors And Consciousness

Our very consciousness is based on a metaphor. All language is a metaphor. The words themselves are metaphors for the things they describe. Not only nouns but also verbs, adjectives and adverbs.

It is difficult to find anything that is not a metaphor. Even a door or window is a metaphor. A chair is a metaphor. You see a chair, window or a door and you know what it is meant to convey without experiencing it. Until you sit in the chair, until you go through the door, until you look through the window, it is simply a metaphor for itself.

---

[141] The further reaches of human nature – Abraham Maslow
[142] These are specific NLP techniques and one of a variety of ways to access stored experiences and states.

## Perceptual Positioning

There are many levels of perceptual positioning. People spend a lot of time in level zero - existing. Sometimes we go to the next level and ask – "Why do I exist?" Rarely we go to the third level – "Why am I asking why do I exist?"

Being able to do this is what allows us to examine our model. Once you know the man behind the curtain is a humbug, your fear disappears.

By standing at this meta position we are not defending our model but examining it. We are examining the motivation for our motivation.

## Gregory Bateson's Learning Levels

Level zero is learning it, level one is learning how you learned it and level two is learning how you learned to learn it. For many people, grasping this is a moment of pure revelation.

## The Matrix Metaphor

It is very like the Matrix[143] moment when Neo realises it is all code and that he can change it. The main thing you do at this third position is to stop defending your model.

Metaphor is a short-cut to dissociation. It must be said that dissociation is a nice place to visit but I would not want to live there. Association is the pay-off; it is the realm of pleasure.

Everything in the universe is described by its relativity to everything else. We need to be constantly reminded that it is

---

143 Motion picture made in 1999 written and directed by Andy and Lana Wachowski. It deals with the nature of reality from the point of view of a character played by Keanu Reeves who discovers that what he and everyone else considers to be reality is a virtual reality known as the Matrix in which all of humanity is imprisoned.

all a relative metaphor. When we are reminded we also get perspective.

## Problem Solving With Metaphors

In terms of problem solving you need to gain as high a perspective as necessary and no higher[144].

The following example may help explain how metaphors allow dissociation from damaged models then re-association with clarified outcomes:

> I was working as an analyst with a company that had computer networks on sites around the UK. They were experiencing serious system faults on an increasing number of sites. The support team were in crisis and under extreme pressure to discover why this was happening.
>
> Each site lost considerable income while the systems were inoperable, running to thousands of pounds per site per hour. A lot of money had been spent installing the systems and fingers were being pointed. The situation was on target to damage several careers.
>
> The IT director asked me to help the support team by creating a program that could analyse the faults on stricken sites. They wanted a diagnostic program that would identify what the fault was, how it had become faulty and the effects of the fault. They reasoned that they could use this information to fix sites more quickly.
>
> I looked at all of the support notes and problems that they were finding. I realised that most of the problems were intermittent and that there had been no clear remedial action or any predictable pattern in which devices were failing. Each fault exhibited different symptoms on different devices. The boundaries of the problem seemed to be the entire system.

---

144 See the metaphor creation tool I described in volume one.

The program they requested would have to be very sophisticated in order to identify and analyse the faults that were cropping up. A team of people were already poring over a backlog of log files and program dumps as sites were falling prey to one problem after another.

The system was one which linked order, delivery and payment in a specialised environment. Each site had a different number of devices on the network. There were three different kinds of device. There were order input devices, printing devices and payment collection devices.

My metaphor had to be pretty literal. I imagined the devices on the network as a group of semaphore towers messaging each other with important information. Something was causing them to ignore critical signals or to repeat them too many times. This was accumulating complexity with time and gumming up the works with complicated instructions. No two sets of towers were failing in the same way. Tower operators tried bravely to keep the signals flowing as the pervasive gum rendered the cogs and gears of the messaging machinery into various states of disrepair. This was making the maps of the territory outdated and incorrect. Incorrect maps of the territory were causing them to think they were getting some messages out of nowhere and to send other messages into a black hole.

The map and the territory were being forced to drift out of true by initial conditions that were either faulty to begin with or that had not adapted to change.

What if I could wash away the accumulated gum and give each tower operator a newly updated and accurate map? What if I could reset the initial conditions?

I designed a program that could be downloaded by the on-site human manager in the event of a site failure. This would include a set of simple instructions about how to uniquely number each node on the network manually with post-its before running the program.

As the program ran on each device it asked the manager how many devices were on the network, which type this device was and which new identifying number he/she had allocated to it. It took less than a minute to reboot and reconfigure each machine with an optimum configuration within its network. The entire network was rebooted when all nodes were cleaned. The resulting configuration for each network was optimal.

I made no attempt to analyse what had gone wrong with individual machines or networks. I was not interested in which nodes were faulty or healthy.

Trying to uncover how they had been configured by analysing symptoms had been making recovery more difficult. It was apparent that some networks had been originally configured incorrectly or upgraded without giving each device a unique identifier or telling the network the correct number of nodes. The problem had a common root but was symptomatically different on each site. Analysing each site, device by device, would only, maybe, provide a solution for that site.

The washing and mapping program worked like a dream. It worked so well on problem sites that it was rolled out as a maintenance task to over 300 sites. Within a month all of the intermittent problems had disappeared and the support team was out of crisis.

Metaphor is an essential tool in this sort of problem solving. You need to dissociate from the problem. A new metaphor can allow you to see where the old metaphor has collapsed.

## Decoupling The Metaphor

Metaphor allows us to step back and step up, across and back down. Most of the time the problem will have dissolved when you get back down to a new metaphor. In many cases this is because the problem was a feature of the old metaphor. Someone had confused the metaphor with reality. They had become worried about problems that the reality did not actually share with the metaphor.

Since almost absolutely everything is a metaphor, and many people think that everything they are dealing with is reality, this is very common.

When the coupling becomes too tight, it is good time to challenge metaphors. When the coupling becomes too tight the map has become confused with the territory and we start to solve problems with the metaphor that have nothing to do with the reality we are dealing with.

Listen carefully and you will detect this when people start to argue about semantics. Semantic arguments indicate that people have started picking holes in the metaphor because they have either lost sight of the reality or they want you to lose sight of the reality.

## Riding The Metaphor

A metaphor is like riding a log downstream. If the log jams you can spend time trying to unjam it or you can jump on a passing unjammed log.

Someone might decide to start arguing semantics and say that when people are riding logs downstream they need to unjam the logs to get them to the paper mill, or wherever they are bound. This is true and it would be relevant if the object of this metaphor was to bind the log to, say, a critical requirement in a project that needed to complete the journey.

The object in this metaphor is not to get the log downstream to the destination; it is to get *you* downstream to your destination. The log represents a metaphor. The metaphor is a means not an end and, as such, is expendable.

So you might decide to change logs. The metaphor now becomes that of the pony express from American history. You are like a rider carrying a saddlebag of critical information.

On a long route the rider changes ponies every time the one being ridden becomes too tired to continue. The spent pony is fed, watered and rested, ready for the return journey while the rider and mail travel on. The object was to get the mail through, not the pony.

Of course this metaphor breaks down again if it is stretched. Metaphors are hugely helpful exactly because they play to our ability to engage in fuzzy logic and to apply judicious deletion, distortion and generalisation in order to make intuitive leaps rather than getting caught in the nitty gritty or the immensity of reality.

## Vertical And Lateral Thinking About Metaphors

Vertical thinking depends on the rigidity of definitions in the same way that mathematics does. It depends on the methodical progression from one proof to the next. It is like stepping from one stone to another across an alligator infested river. You need to make sure the next rock is a rock and not a sleeping alligator. Jumping sideways is rarely a good idea.

The problem is that if you start at the wrong place with some flawed axioms, a careful progression of logical steps will not get you where you want to go.

Lateral thinking allows you to leap sideways onto a passing log in the full certainty that all the logs and all the alligators are imaginary in an imaginary river that you control. When vertical thinking reaches an impasse, lateral thinking pretty much always provides a free pass.

We have come a full circle back to Gödel and the incompleteness theorem that so worried mathematics.

Unjamming the log, as with Russell's paradox, is a matter of levels of classification and perspective. You have to keep climbing that tree in order to take a good look at the territory to make sure that:

1. You are where the map is telling you that you are.
2. The map is still accurate and following it is still taking you where you want to go.

Both vertical and lateral thinking are necessary. They are both tools and you do not get to use one to the total exclusion of the other.

## Semantics

Avoid semantic arguments. Semantic arguments are about the meaning of words and have rarely anything to do with material facts. All words are metaphors. People who know what is meant but continue to argue possible semantic meanings usually have a secondary gain or motivation for doing so. It is sometimes, but very rarely, for clarification. Semantic arguments rarely produce clarity.

Semantic clarification often comes in the guise of the king and emperor of questions, "How?"[145] and "What would happen if?[146]". This is the equivalent of someone handing you a pole with which to either unjam the blockage or vault to safety. Semantic argument on the other hand is usually an attempt to tie you to the log and push it over a waterfall.

## Everything Is A Metaphor

When you find yourself in a situation where an intractable problem has crashed into your path, keep this in mind.

Let's take project management or software design. We often hear we can't do this and we can't change that. This is merely a metaphor collision. It is caused because people have a model of what project management or software should look like. This model is their metaphor for dealing with something

---
[145] How will that help us? How is that a problem? How will we deal with X? How will that affect y? How many do we need? How would that apply to Z?
[146] What would happen if we were to do that? What if we did not include that? What if needs to be followed by another how question.

very complex. In their model there are reports, alliances, ceremonies, dependencies and assumptions that have hardened into the simulacrum of the bone of fact. It's all just a metaphor. It crumbles when it is of no more use.

> "You do look, my son, in a mov'd sort,
> As if you were dismay'd: be cheerful, sir:
> Our revels now are ended. These our actors,
> As I foretold you, were all spirits and
> Are melted into air, into thin air:
> And, like the baseless fabric of this vision,
> The cloud-capp'd towers, the gorgeous palaces,
> The solemn temples, the great globe itself,
> Yea, all which it inherit, shall dissolve
> And, like this insubstantial pageant faded,
> Leave not a rack behind. We are such stuff
> As dreams are made on, and our little life
> Is rounded with a sleep."
>
> – Prospero in act 4 of Shakespeare's The Tempest

There is no such thing as an intractable problem. When a problem does seem intractable, the real problem is often that people cannot see a solution in a metaphor they refuse to release. If we look at many world events this seems to be true e.g. In South Africa a problem dissolved when Nelson Mandela gave the country a new metaphor and so the Rainbow Nation replaced Apartheid.

We are such stuff as dreams are made on.

# Concentric Contexts

*"He who loves practice without theory is like the sailor who boards ship without a rudder and compass and never knows where he may cast"*
**Leonardo Da Vinci**

## Knowledge Is Infinite

Pick any strand of enquiry and pull it. It will lead you on to infinity. This is the nature of reality.

There is an infinity of infinities. Some infinities contain each other.

The set of prime numbers is infinite, (2,3,5,7...) but it is contained in the infinite set of natural numbers (1,2,3,4,5,6,7...).

Infinity is probably a metaphor. In a certain light that would make existence a metaphor. In a certain light we could argue ourselves out of existence, so we cling to our pitons of logic.

Mathematics is a strand of knowledge. It is a particularly powerful one. If we consider it as a circle, it is an infinity that contains an infinity of infinities. It is bounded by a system of logic. This is also infinite. This too is bounded by a system of science. Science means knowledge. Knowledge is an infinity of infinities.

## Enclosing Levels

Every circle of knowledge is surrounded by another containing system of knowledge. The boundaries of every circle of knowledge are created by the rules of the next level.

This is what Russell found when he hit the boundaries of set theory with his first throw.

From within any circle you consider rules from anything outside to be illogical, irrelevant and possibly immoral and insane. While you are outside the contained circle you consider the rules inside to be inadequate and short sighted.

Thinkers and philosophers have chased Reality throughout the ages. To catch a glimpse of reality they had to let go of all limiting beliefs about themselves and the world. This can mean anything we cling to as a sense of self. We may even admit that whatever we consider to be reality can only be a perception, made up of our fears, hopes, prejudices, beliefs, and very limited point of view.

## Reality Keeps Changing With Perspective

We know that the physical and mental horizons of humanity have continued to expand. We know that as we develop tools and intellects that allow us to look at the depths of interstellar space and the building blocks of life, we have had to shape and reshape our ideas about ourselves and our place in the Universe.

Imagine that we can see a reality that can contain all of the models of all of the consciousnesses that have ever existed.

Some of those realities are very limited and exist within the shells of almost static energy levels. Some of the realities are complex and expanded by knowledge that approaches huge scope and complexity.

We can recognise that we have a hierarchy of needs

Clare Graves was a student of Maslow and his theory builds on Maslow's hierarchy of needs. Graves discovered that people, communities and civilisations evolve along similar lines. He identified levels of human existence[147] and colour coded them in order to disentangle them from any ideas of good and evil or relative worth.

---

[147] His paper "*Levels of Human Existence: An Open System Theory of Values*," was published in the Journal of Humanistic Psychology in 1970

## The Graves Levels

- Beige – Survivor: Whatever it takes to survive.

- Purple – Tribal: self is subsumed into customs and culture:

- Red – Egocentric: Get whatever you want through power.

- Blue – Authoritarian: Obedience to rightful authority.

- Orange – Scientific: Clever individuals achieve progress.

- Green – Harmony: sacrifice self interest for acceptance.

- Yellow – Systemic: Express self and avoid harm to others.

- Turquoise – Holistic: Accepting existential dichotomies.

- Coral – Transcendent[148]

## Ubiquitous Levels

Every level of perspective is defined and enclosed by another level. This idea surfaces everywhere there is progress, investigation and knowledge. It is process of the discovery of new levels of perspective. Consider how Russell and Whitehead's attempt, to prove mathematics, was put into perspective by Gödel's surrounding incompleteness theorems. Consider how Maslow's hierarchy of needs, complete in itself, was enveloped and enhanced by the findings of Clare Graves. Each stepped to another level to provide deeper understanding and open the aperture of knowledge. We are constantly witnessing feedback loops in the recursive engine of reality.

In the ideas of perceptual positioning or the basics of negotiation we see, over and over again, a pattern that defines reality. It is always possible to abstract to the next level or

---

[148] "Timeline Therapy and the basis of Personality" by Wyatt Woodsmall and Tad James contains a more complete explanation of these levels.

instantiate to the previous level. Whatever level you are at becomes reality. It is a function of the brain to adapt.

Whatever the amount of information we are dealing with, we always get to a stage when we have to wrap it up in something in order to manage it.

We went from a flat world to a round one to one that orbited the sun to a sun that orbited a galaxy to a galaxy that orbits the galactic core. We went from standing up to gravity to relativity to quantum to string and beyond. We went from three dimensions to spacetime to quantum dimensions to holographic theory. We found the atom, then the proton, neutron and electron, then to quarks and god particles and we are searching for super symmetry. Each stage has an anomaly that opens a door to a bigger infinity.

I predict that whenever we get close to any unifying theory we will find one little problem that will stump us, puzzle us and confound us. It will be this problem: The unifying theory will not be able to contain itself. We will find that we jump to the next level where we have a unifying theory of unifying theories and the pattern will repeat.

Reality is both fractal and recursive.

# Methodological Conclusion

> *"Begin thus from the first act, and proceed; and, in conclusion, at the ill which thou hast done, be troubled, and rejoice for the good."*
> **Pythagoras**

## TOR – Terms Of Reference

A definition of a project is the organisation of resources to achieve an outcome. By this definition it is possible to consider almost everything we do as a project.

When we are successful at something we try to repeat that success and we develop strategies for doing it. If certain important and key aspects of the situations to which we apply the strategy are similar enough to the key aspects of the original situation in which the strategy was successful, the strategies tend to work.

On the other hand, if we apply the strategies based upon superficial aspects of the initial conditions, then we have abandoned ourselves to luck, probably failure and certainly hardship. Even though we think we are being terribly scientific and pragmatic, we may not be following the recipe for success we imagine.

## Terms Of Reference Are Initial Conditions

### *We identify*

- ✔ Background
- ✔ Objectives
- ✔ Scope
- ✔ Constraints
- ✔ Assumptions
- ✔ Reporting
- ✔ Dependencies
- ✔ Estimates
- ✔ Timescales

### *We ask*

- ✔ What are the purposes, aims and deliverables for the project?
- ✔ Where are the parameters in terms of time, money, authority and scope?
- ✔ Who is involved, how will they be organised, who will make decisions?
- ✔ How and when will progress be measured?

## Methodologies And Myths

Methodologies are simply groups of strategies and the techniques used to realise them. They are a surface structure.

The world of methodologies can be confused and confusing. Many things have been mislabelled in an attempt to commoditise and package fairly basic behavioural strategies. The confusion is that these methodologies are often tactics packaged as strategy. They deal with initial conditions as if they were the deep structure of the project. The deep structure of the project emerges as the project is implemented; primary and secondary gains are identified and prioritised; and reality is bombarded with particles.

## The Three Levels Of Focus

We consider time-scales, resources and content. We measure quantity, cost and quality. We drive these things forward ever mindful of risks and constraints. The methodology or behavioural strategy is the biggest risk and the most limiting constraint unless there is context.

Everything we do has at least three levels. How, what and why. This contains the skill, the vision and the motivation. Very often the motivation seems obvious but there is almost always a secondary gain. This secondary gain is almost always unspoken, untracked, unrecognised and, more often than not, unconscious.

It is easy to become obsessed with **how** to do things. We think we are in control when we determine **what** we are doing. Real control comes from knowing **why**.

To do this we have to dispense with the answers that will please the shareholders and the interview board; the crowd in the pub; and the purveyors of closed questioning. To discover why and to learn how to uncover secondary gain, we sometimes have to prod some very sacred cows with some very sharp sticks. Some of them explode. Messily. So we have to watch our language carefully as we go.

## The Most Sacred Of Sacred Cows

### *The place of business in society*

Is money really the reason businesses exist? Is profit really the motivation for all human behaviour?

To listen to some captains of industry expound, it would appear that society was invented to support business and the making of money. It would seem clear that profit is the ultimate goal of human consciousness and existence.

### *Prodding gently (How)*

If we look at the roots of business, the word comes from the old Northumbrian word bisignes, meaning care, anxiety and occupation. We carry out business in companies, a word based on the Latin word companio, meaning companionship, society and friendship. We set up these companies within industries, which comes from the Latin words indu, within, and struere, to build. These are all very positive and society friendly ideas.

### Prodding more vigorously (What)

One does not have to be a historian to imagine the roots of business in every culture on earth. People got together to do things they could not achieve alone. People supported and rewarded farmers, artisans and suppliers whose efforts allowed resources and skills to benefit all. Money was quite probably invented because barter relied on the supplier of "what you want" having a use or need for "what you supply". This double coincidence of wants[149] is unlikely, so we needed a way to store and apply credit for our contribution outside of immediate transactions.

### Prodding with gusto (Why)

Looked at this way, it seems more likely that commerce was invented to support society and that profit was a by product of pragmatism.

Society is changing around us and people are more educated than ever before. People in Western society see life less as a struggle to survive and to amass wealth for its own sake and more as a brief opportunity to find some meaning in their own existence. People are becoming more aware of the quality of life.

In this context the definition of success and failure is constantly changing. Measuring it solely as a stark financial outcome verges on a compulsive disorder. Many companies now realise that their worth is not merely financial, but societal. Successful companies are quite often run by very intelligent people, particularly companies that are required to be creative and productive in order to succeed. These people have access to the same moral loadstones as the rest of us,

---

149 Double Coincidence of Wants – William Stanley Jevons (1835-1882) mathematician and logician coined this. He also coined the Jevons Paradox: he deduced from his study of coal usage that improvements in fuel efficiency tend to increase the rate of usage of that resource rather than reduce it.

and, unless they are psychopaths, they tend to react with more humanity than we often credit them with[150].

## Business Relies On Society

Business may be the engine room of "the good ship Society" but the engine relies on the ship to keep it afloat more than the ship relies on the engine for thrust.

It seems redundant to have to state it, but if society collapses, so to does industry, commerce and business. It is in their own interests to contribute to the advancement of society. A surprising number of the successful business people realise this very well. It appears that the psychopathic idiots, intent on drilling a hole in the bottom of the boat, are only a noisy minority. This does not stop them having a disastrous effect.

We can see evidence everywhere of even massive companies, whose financial successes are so huge as to be almost beyond understanding, striving to be of value to society beyond being a hoarder of wealth. The best ones appear to crave meaning within the context of something wider than the spreadsheet.

### *Time is money, money is energy*

Money was once described to me as an energy like any other energy. It has huge potential to create or destroy. It is another double headed axe.

Finance is a surface structure. It has become distorted in many minds but it can be a useful tool. Like any tool, when we

---

[150] Whether you admire or despise someone like Bill Gates for instance, there is no denying that he does not have to spend his money and his time in pursuit of philanthropy. He is rich enough to found his own country, if he wanted to, and to surround himself with a reality devoid of unpleasantness. He does not need our approval and is surely intelligent enough to know that whatever he does his motivation will be questioned and some people will always despise him. Yet he spends his time trying to make a difference in very unpleasant situations. I am sure he is no saint but he is just one example of a ridiculously wealthy person who seems to seek success other than profit.

treat it is a means and not an end, it becomes easier to deal with. There are also phobias associated with finance. Finance hides many secondary gains. How to deal with these is explored in detail in volume three.

## Denominalising The Sacred Cow

We can move the discussion of modern life to a context where we identify our lives as part of the continuum of human history. To do so we have to denominalise all the labels that pose as reasons and replace the labels with real values – in the same way we would solve a quadratic equation by finding values for X and Y.

## Denominalising "What"

If we ask people what they are doing, they usually start to tell you how they are doing it.

Examples:

> *Q: What are you doing?*
>
>> *A: I am mixing some cement to pave the garden.*
>
> *Q: What are you doing?*
>
>> *A: I am setting up this machine as a server to continuously merge compiled and tested code.*
>
> *Q: What are you doing?*
>
>> *A: I would have thought that was obvious, I am hammering nails into this wall.*

## Denominalising "Why"

Question why they are doing it and you will elicit what they are doing:

> *Q: Why are you doing that?*

> **A:** *I want to have somewhere to park my car.*
>
> **A:** *I want to have a fully automated continuous integration suite.*
>
> **A:** *I want to hang up some pictures.*

## Denominalising "How"

Here comes a piece of strange recursion. It is one of those strange loops. If you want to get to level three, when asking "What?" gets you "how"; and asking "Why?" gets you "what"; what do you imagine you ask to get "why"[151]? You ask "How?" of course.

> **Q:** *How is that going to benefit you?*
>
>> **A:** *I am afraid my car will get stolen if I park it on the street. Also, my insurance premium is lower if I park my car in my drive.*
>>
>>> We are also interested in the 'secondary gain' – in this case it could be that if there is less grass to cut, I will be able to spend Saturday afternoon watching football.
>>
>> **A:** *I am afraid that if I leave it all until the end of the project, that integration will be a huge task and some of the changes will interfere with each other. If we integrate all the functionality as it is added we can make minor adjustments now and that* will *save us lots of time at the end of the project.*
>>
>>> The secondary gain could be that we always demonstrate progress by allowing our customer and our project manager to play with the latest working functionality so they will trust us and make the project more stress free and fun to work on.

---

[151] With apologies to Abott and Costello "Who's on first base".

> **A:** *I will have some images in my line of sight that will make me feel more comfortable in this room. There are holes in the plaster and I am covering them up so I don't have to repaint or re-plaster the wall.*
>
>> A secondary gain could be that if I have my pictures up, I am demonstrating that I have good taste. Potential partners will find me interesting.

### *The secondary gains here are all guesses.*

Some schools of psychiatry claim to be able to always ascribe the same motives to patterns of behaviour. I find this dubious.

Patterns of behaviour without context tell us very little. What may appear to be destructive or harmful behaviour in one context, can be resourceful or helpful behaviour in another. Defining this context is what project management is really all about.

If you rely on techniques or methodologies for context, you are already lost. They have their own context already built in and they rarely care about yours.

# The Evolution Of A Crime Against Reality

## *Introducing the methodology*

Traditional waterfall methodologies present the world like this:

| Big Picture | why |
| :---: | :---: |
| ⇩ | ⇩ |
| Methodology | what |
| ⇩ | ⇩ |
| Tasks/Detail | how |

This leaves an embarrassing "turtles all the way down" scenario for the big picture. It is supposed to arrive fully formed and complete. We have already discussed at length how reality laughs in the face of this. Therefore a strange loop is constructed.

## The rise of the methodology

Often the methodology determines everything and you are encouraged to see it as the project constant. "Why" and "how" are subsets of, and must fit within, "what" (the methodology). This is a boundary masquerading as a constraint.

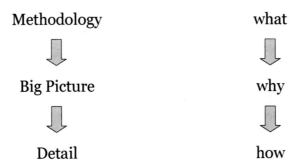

This still leaves "how" with some room to manoeuvre. While "why" is artificially constrained, its very nature is to be detail-free and open. Since practical skill tends to grow and adapt with the people practising it and the advances in its professional groups, some skills cover more territory than the methodology or even a specific "why". Methodologies are invariably hindsight rather than foresight. This leads some methodologies to posture as foresight and to turn the telescope right around in an attempt to neuter reality.

> In a software project the requirements are constrained by the methodology. The computer code must still implement these requirements whether they are relevant or not. This often leads programmers to side step the boundaries and shortcomings of the project methodology and to deliver functionality in creative and unpredictable ways within the requirements. They can also embarrassingly often use fact based arguments to highlight when the requirements boundaries are too far divorced from the realities of the technology or real world constraints.

## *Customising reality to fit the methodology*

For those who sell or champion methodologies, the secondary gain is that reality gets tidied up. It has its wavelength topped and tailed, like a digital recording, so that it fits neatly into the methodology. "What" (the methodology) deletes the links between "why" and "how" and sets its own boundaries for both "why" and "how".

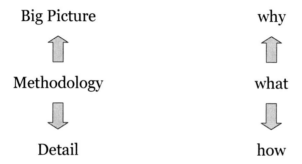

With "how" under control there is a conspicuous opportunity for efficiency. "What" no longer needs to control "why", it can leave a neutered "how" to take the blame for that.

> In a software project the methodology can easily constrain the technology available by introducing process requirements, arbitrary boundaries and technical decisions at a management level without getting real programmers (the people who are going to have to do it) involved. This makes it appear that there is a shortcoming in either the technical skills available or the ability of the technology to deliver. Details are decided upon too early and when the time comes to discuss details later in the project, the people with authority normally retreat to the security of the big picture behind a wall of methodology and process. It is now brilliantly easy to shift the focus onto "how" or "capability of subordinates". When the programmers try to explain, they can easily be considered out of touch with the big picture, which of course they are, because they are on the wrong side of the methodology wall.

## *Reality demoted*

When this happens, and the methodology becomes the rule and everything inverts. Reality is viewed through the detail of the methodology. The methodology becomes the only filter and the main constraint. However what happens when the gamekeeper is also the poacher? The methodology will never find itself to be the problem.

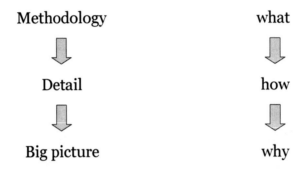

## *Reality fights back*

Reality is big and loud and messy. It rarely restricts itself to only one dimension let alone one direction.

Observation, experiment and experience all teach us that, no matter what we would like it to be, reality has its own ideas about cause and effect.

Sometimes we only understand why after we have begun.

## The methodology as a servant

The methodology must be the servant of what we want to do and the abilities and potential abilities we have available to do it. "What" must change to accommodate "why" and "how". The methodology must be a feedback mechanism not a self referencing filter. As we discover new needs and new abilities a good methodology will have the facility to be ego-less and to adapt to our changing map of reality.

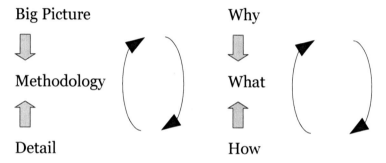

In neuroscience the paradox is that what we do can limit our capabilities or stretch them and that our capabilities, driven by our nature, continuously change to challenge or limit what we do. Our needs and our capabilities determine what we do. They are our natural drivers. If we allow them to grow by freeing "what" from boundaries masquerading as constraint, what we can achieve evolves and expands our boundaries and our horizons. How often have you heard someone with great talent declare that they have no choice about following their heart? This is another way of saying that they have to explore and use their talents and abilities in order to find fulfilment.

> In a software project approaches such as Agile ideally adopt an adaptive cycle of feedback and learning in order to create organisational and personal growth while achieving realistic and valuable goals.

### Motive and outcome entanglement

This recursion is recursive again. As we learn about neuroscience, it has become clear that nothing is fixed and that the way we do things begins to affect us in unexpected ways. When we do something habitually, it becomes our capability. Motive and outcome become entangled.

What starts to limit or enable why and how.

### Liberating why and how from what

Agile and NLP both look very like the final diagram above. They seek to step up into the "why" level and they recognise that capabilities are constraints that must be developed with purpose.

Nothing can be done unless we drive some pitons into the wall of reality and start climbing. As I heard the astrophysicist Neil deGrasse Tyson comment in a recent interview[152] "When you are on the frontier of knowledge between what is known and unknown, reaching out into that abyss, sometimes you do actually have to make stuff up that might be true, so that you can organise a research plan to find out whether or not it is."

So, when we find ourselves on the frontiers of knowledge....we make our best guess, based on what we think we know.

## The Frontier Of Knowledge

Most of us are not at the frontier of knowledge. That frontier has moved way ahead of us. The frontier of knowledge has moved on and it is our job to stop arguing semantics. We need to open our eyes and ears and pick up the trail. Many of us are settlers arguing about the positioning of fences, while the real pioneers are disappearing into the distance. It is time to recognise that there is an ever expanding horizon and that we

---

152 The Daily Show January 18th 2011

are wasting time squabbling over the price of a few shiny pebbles that have distracted us while we could have the stars.

## Levels Of Elevation

A sextant is an instrument that was used to navigate at sea. By holding it vertically, navigators could measure the angle between a celestial object and the horizon. That angle was known as the altitude or the elevation and allowed them to calculate their latitude. The principle of a sextant is that it allows measurement of the angle between *any* two visible objects. Turning it horizontally would allow them to measure the angle between landmarks and use triangulation to calculate their position on a chart.

Navigating a model of reality, such as a plan, means being able to calculate your position on the model. This is not enough. We need to know the position of the model with reference to landmarks of reality. We can use our combination of vertical and horizontal thinking to check and adjust our position.

In other words we can continue the tradition of using elevations to gain levels of perspective. In our concentric energy levels of reality, our elevation tells us what level we are dealing with.

### *Methodological Elevations*

1. On level one there are groups of techniques or individual methodologies
2. On level two there are groups of methodologies

To many of us this level is the level at which we justify what we are doing and consider that we are managing the situation. We talk about targets and deliveries.

3. On level three there is the method for methodological approach. The principle here is to apply the spirit of the

law rather than the letter of the law. It is a matter of abstraction.

## *Abstraction*

These levels of elevation always exist in the same reality although sometimes they seem unconnected. The more angles of perspective we have the more connections we can make. As leaders and managers it is not enough that something is working, we need to know why it is working and how to extract the principle. We need to be able to apply principles as techniques and to extract principles from working techniques.

1. Filter: Why are we thinking of using this technique or tool? What conditions does it specifically require to function properly? How can we make it more self sufficient by adding or taking away dependencies?

2. Intersect: Why does this approach work? What does this have in common with other tools and techniques that work? How specifically was it designed to work? How will we know that this tool or technique worked or is working?

3. Connect: Why do we consider this success a success? What other ways could we succeed? How can we use the principle without the technique?

## *Levels*

It is important that, as we abstract levels of doing, we can navigate them and find our elevation. This law of three levels can apply to just about anything, and as managers we need to recognise and find our way to level three.

1. Do something repeatable

2. Make an environment for doing it again

3. Make an environment that generates environments

# Framing Up

> *"A purpose of human life,*
> *no matter who is controlling it,*
> *is to love whoever is around to be loved."*
> **Kurt Vonnegut**

## 13 Billion Years, 10,000 Galaxies

The world becomes, day by day, a more fascinating place. The scale of the universe and the mysteries of the mind humble us. They make us want to celebrate the act of existence while reminding us of the speed of our flight through the lit room[153].

Despite all the uncertainty we face about the future, one cannot help but to be lifted out of a pedestrian view of reality when faced with a video[154] that talks about the universe being 93 billion light years across.

You have to love the human brain and be delighted when you find arguments underneath it on Youtube disputing the accuracy of 93 billion light years. There are wonderful objections stating that the farthest we can see is 13 billion light years and passionate disputes about the speed of light as if we were discussing train times.

There are images[155] available at the touch of a button that take us on a virtual tour of the known universe based on the current state of the science of astronomy.

---

153 An advisor to King Edwin of Northumbria is said to have described life as the swift flight of a sparrow through a banqueting hall. After the brief comfort it is back outside to storms and darkness. We do not know what went before and we do not know what comes after. We only know the warmth and light as we fly through.

154 http://www.youtube.com/watch?v=e7t5d0eseIs
(You are here)

155 http://www.youtube.com/watch?v=17jymDn0W6U
(the known universe from the american museum of natural history)

There are other videos[156] in which there are moments when only the most jaded heart would not skip a beat. We are told about a region of blank space in the night sky, smaller than a grain of sand held at arms length. That region is the furthest we have ever seen, and we see it with light that took thirteen billion years to reach us.

In 1996 the Hubble telescope saw, in this tiny region of the sky, three thousand galaxies each containing hundreds of billions of stars.

In 2004 it looked at another even further region, now known as the Ultra Deep Field, and saw 10,000 galaxies.

In the glory of the night sky this region of sky had been nothing but a small dark speck. Now we can see ten thousand galaxies there.

Which is more impressive, the scale that this gives the universe or the ingenuity of our brains? It can make you feel tiny in such a large universe or it can make you feel that your mind has travelled so very far, so very fast to connect with something that has existed for so very long.

## Accelerated Through Our Own Brain

We bounce like particles in an accelerator through the brain. On our way we strike neuroscience, chaos theory, evolution theory, information theory, quantum theory, music theory, history, physics, chemistry, biology, art, music, architecture, economics, technology, methodologies and techniques of all sorts, and glance off untold others. In doing this we dislodge questions, like atomic particles, that show us the deep structure of the possible.

---

156 http://www.youtube.com/watch?v=oAVjF_7ensg
(the ultra deep field in 3D)

## Communication Is Feedback.

We are a matter of communication and feedback. Communication is two-way. We transmit and we receive. We respond to feedback with feedback. We are like our very own Mandelbrot image as we break life down to the self similar. We absorb the feedback and shape it so that we can understand it.

## Data Compression

Information theory provides the theoretical framework for data compression. Data compression is used for optimisation. We attempt to maximise the productivity of the messages we can handle and minimise the cost of expensive resources by making less mean more.

Machine learning, or AI, uses information theory and data compression. The theory is that as the computer experiences and learns, its history represents a familiarity that allows it to accurately predict more of the message from less information. This allows compression to gradually increase so that the size of the data transfer during the interaction decreases and the speed increases.

Every technique can be refined to some sort of interaction or transfer of information. We interact with others but primarily we interact with ourself. We undergo a similar process of data compression.

Information becomes data becomes knowledge becomes adopted habit (or wisdom). This happens by removing blocks and constraints to communication and replacing them with filters that allow us to recognise the shape of the messages we send and receive and guess at the *a posteriori* possibilities (i.e. presuppositions).

Effectiveness is doing the right thing and efficiency is doing it as well as you can. It is recursive. We increase the gain, we clarify the signal, we increase the information content.

Habit is data compression. The first time we do something we analyse all the information and much of it is irrelevant. As we practice and get familiar, not only do we filter out the irrelevant information but we start to predict. We can get very, very good at this and we call it skill. We get a lot more feedback and useful information from less and less data travelling across our bandwidth. This happens because the neurons in our brain actively rearrange to specialise and look for specific information.

> A finger pressing a piano keyboard for the first time feels either press or don't press. It is for and against. It is *either-or*. As the pianist progresses in skill, the neurons controlling the finger tip increase and each one compresses the message to the brain to carry specific and meaningful data. The brain decompresses these, delivering nuances of pressure that can make the difference between talent, skill and genius. Consider the fingertips of Chopin or Arthur Rubenstein in relation someone taking their first lesson.

As we develop habit, we increase the sensitivity to feedback. We increase the distinctions we can perceive. Perception: the same action of the finger of a trained musician has compressed a universe of information that the musician does not even consciously think about.

Simple things can be scaled up to complex functionality. They optimise and carry more and more information with them – weightless, priceless information.

When the signal is weak the receptor compensates, when the receptor is weak the signal compensates. When the signal gets through there is action.

Communication is how we assess the situation, assess our abilities and assess our understanding of our will. It relies on sensitivity, accuracy and synchronisation at the signal broadcaster and the signal receptor.

When this goes wrong and communication is confused we perceive problems. When we mismatch the model and the world, the map and the territory, the design and the ability, the question and the comprehension, the weave and the weft things get out of balance. When balance is lost, things jam.

## Approximation

We have a ghost in the machine that saves us. We have evolved for the approximation of chaos. We can fill in the blanks. This is what we have done in communicating with ourselves and others. Objectives and outcomes become reality. Communicating causes things to happen.

Outcomes are what we tell ourselves we want. We believe ourselves so we had better be sure that is what we really do want. When you know what you want you are most of the way to achieving it. Many of us have a fog of uncertainty in our brain. Communication is about clearing that fog.

This volume is about what to do. What you need to do is to keep discovering what you really want to achieve. To do this you must check your model against reality. Volume three examines some useful techniques which enable you to progress from the "what" to the "how" to achieve your goals.

## The Mirror In The Mirror

To communicate with others you must communicate with yourself. To communicate with yourself you must communicate with others.

Hegel said that "culture means nothing else than that this substance gives itself its own self-consciousness, brings about its own inherent process and its own reflection into self."

This is the mirror in the mirror. It is the ultimate recursion. Consciousness is the infinite recursion of self-consciousness recognising self-consciousness.

Love is a recursive mirror. The point of being with each other is to provide a feedback mirror to perfect self awareness. Self awareness mandates denominalisation of the model. The model is not a thing, it is a process. There are consequences to allowing the model to set. It is very much like allowing the body to become muscle-bound. Muscles are meant to move and stretch to provide movement.

The point of collaboration is to provide a feedback mirror for each other and for ourselves. This keeps the model in flow.

Humanity's model is changing all the time – do you really think that you can sit there and hold on to your tiny model of the world while we as a species are racing ahead? The model is a tool – it is not a permanent structure – you are only cheating yourself if you try to hang on to an after image of your own consciousness.

As we gaze out into the universe we mirror it. We mirror it within our model of it. This is a two way mirror allowing reality to model us. We become what we do.

## We Send Ourselves

The brain mirrors itself. This is the deep principle. It can look at itself and it can see itself reflected in others like it.

This can be a constraint or a recursive truth. Feedback arises from feedback. When we look at one other we know that one of these amazing mirrors is in front of us. We know that we are not alone and it is only in that mirror that we can see ourselves.

## The Practical Nature Of Love.

Love is when you start seeing your reflection in another's reflection of your reflection. Love is a feedback loop. The clearer the mirror, the deeper the love. This is the strangest loop of all.

People who are in love fall in love with the whole world. They can start seeing in the whole world as their mirror. It can be intoxicating.

This is more than a vanity mirror. Love can shock you out of your habitual filters so that for an instant you start seeing the world unfiltered.

The mirror is a metaphor. When you mirror, you become. You are what you mirror. Empathy raises everybody's game.

## Love And Project Management

Often in society we conflate scarcity with value, and availability with worthlessness. These are limiting beliefs.

There is a sunset somewhere every second of every day somewhere on our planet. This has gone on for millions of years – does that make them any less beautiful? There are six billion brains on the earth; does that make any one of them any less fascinating?

There have been love poems as long as there has been writing. Four thousand years ago they were carved into rock.

Love has been sentimentalised and commoditised for marketing purposes. Does mocking love make it any less the most intrinsically valuable experience we can have? Love is an intrinsic part of being human and we must never dismiss it.

I could be talking about romantic love but in this context I am not. I am talking about love as the recognition that another person has the same complexity and inner life as ourself. It is a theory of mind and a theory of love. It is the empathy that all at once expands the infinite universe inside our head and expands it to accommodate every person we meet. It gives purpose to communication. It is the recognition of ourselves in the mirror of another person, knowing that they recognise themselves in us.

## Heaven Or Hell

You can choose to go with Sartre and declare that "hell is other people." It can be hell, especially if you do not pay attention to communication and the mirror gets warped. That may have been what Sartre meant too. He claimed that this quotation was frequently misunderstood.

Like it or not, we need other people and we mirror them. We bring this mirror with us when we encounter other people. It is part of the human condition. We control our reality through our thoughts. In a business sense this takes the form of respect and professional acknowledgement. We must realise that we are interacting with fellow travellers.

The most amazing, switched on person and the most stupid, unpleasant person you met today had this infinitely magnificent thinking mirror in their head. They can all choose to use plasticity to imprison themselves in routine and limiting beliefs or to open new horizons and levels of experience.

## The Inescapable Conclusion

There are currently nearly seven billion of these mirrors in the world. That is seven billion minds regarding themselves and regarding the universe around them. That is around 6% of all the people who have ever lived. This number is growing exponentially. In 1810 there was a population of one billion people in the world. In 1910 it was just under two billion. In 2010, despite two world wars and astonishing progress in weapons technology in the previous hundred years, there were approaching seven billion human minds on this planet.

That is a lot of models. If you believe in control rather than exploration, it is a big problem. It is a problem when we perpetuate the ideology of zero sum and king of the castle games. In doing so we attempt to turn the losers into beasts of burden and we extinguish the minds that could hold the answers to the practical problems of energy, food and how we can find a way to balance our interactions with our environment in order to enjoy it without destroying it.

These games lead some people, especially in the industrialised world, to spend much of their time and effort trying to deaden their perceptions. Like Rilke's panther, they pace in circles, convincing their brains that the circles are prisons or fortresses, lead or glass. The longer they trudge, the deeper the plastic rut in their brain becomes.

### *Notes on process over state*

Balance is a process. Something must be able to move. The walker on a tight rope must be able to move the centre of balance to keep it over the rope as progress is made. The walker on a loose rope must be able to move the rope to keep it under the centre of balance as it moves forward. The power in a fulcrum is that it can be moved to allow limited work to lift different loads. The process must be kept agile so that it can be balanced.

The balance of knowledge is not between good and bad. Although the splitting of the atom meant also the discovery of the atom bomb, that is not the balance. The mirror of versatility is not the balance between good and bad uses. The mirror of versatility is the knowledge of the full range of movement and the potential uses of knowledge and how we interpret them.

The balance is responsibility and that is about understanding consequences. There is a difference between ecology and ethics. Ethics are a set of justifications based on models of the world. Ecology makes us aware of the consequences and asks us to consider them and whether we should accept them or reject them no matter how justified our action is ethically.

## Polishing The Mirror To See A Better Reflection

> In the story of Narcissus, he was so captivated by his own reflection in a lake that he starved because he could not bear to look away. There is another tale told from the perspective of the lake. It was asked why it had captivated him. The lake replied that it had not been interested in Narcissus but in its own reflection in his eyes.

For anyone to truly hear, they must first of all hear what they are saying themselves. To lead, to coach, to manage, to inspire you must enable others to communicate with themselves and to hear themselves in what you are saying. You must help them realise their resources and help them focus on the outcome.

### *To do this:*

Train the unconscious mind, reveal resources and get rid of the time wasting rationalisation and blame shifting. Hear the signal through the noise of the energy wasting messages of ego and excuse. Waive your right to pro and contra, they are noise.

Communication on every level is a shared message. It is not the loudest, but the clearest message that should be listened to.

Arguing about the style confuses the message and the intent of the message. Discord can be useful in dislodging assumptions while harmony gives direction. In NLP terms mismatchers[157] can dislodge complacency while matchers[158] can encourage collaboration.

NLP teaches us techniques that enable us to communicate better by reading all the information in the signal. It teaches us to notice and match subtle distinctions in language, timing, physiology, pace and emotional state. In other words we

---

157 People who look for differences.
158 People who look for similarities.

optimise our communication by understanding human compression techniques.

## Balance, Predictability And Resolution

Chaos does not bring predictability, but it does bring opportunity. Chaos is the way of reality. You must ride it with your senses on fire as it loops and spins bringing the future and the past into focus through the present. You can only cut a straight line when you have surfed into the heart of being and realised that it is all so much bigger and so much smaller than you had imagined.

The balance is the tension. The journey is not only more important than the destination, it is the destination. You must keep going. To stop is to stop creating, to stop creating is to stop existing.

> Many years ago I took a walk in the Australian desert with an Aboriginal man. He pointed to a painting on a cave wall. This is our most revered and ancient painting – he told me. I examined the painting. It was beautiful but I expressed naive amazement that such an old painting could be so vibrant. He laughed and told me that I misunderstood. The first painting had inspired the next painter to paint on the same spot and that painting had inspired the next on the same spot for thousands of years.
>
> I was confused. In my western model this was desecration. He shook his head. You still don't understand – he said. This is respect. We respect the creator more than the creation. We respect the act of creation above all else. We honour the creators that have created here by partaking in the same act with them.

To be alive is to have a consciousness that is aware of creating. Anything is possible, a butterfly and a storm come from the same simple rules. What are the odds against existence in this time at this place in the immensity of time and the immensity of space? Yet we prove the impossible by existing in this time at this place. Why limit the possibilities to gods and monsters?

*"The key question isn't "What fosters creativity?" But it is why in God's name isn't everyone creative? Where was the human potential lost? How was it crippled? I think therefore a good question might be not why do people create? But why do people not create or innovate? We have got to abandon that sense of amazement in the face of creativity, as if it were a miracle if anybody created anything."*

**Abraham Maslow**


The universe is movement, it is pent up energy, it is entropy. Even in stillness we resonate to the beat of the stars. "Eppuar Si Mouve". Reality moves.

# Consequences

*"It is not enough to have a good mind, the main thing is to use it well."*
**Rene Descartes**
*"Intelligence is something we are born with. Thinking is a skill that must be learned."*
**Edward de Bono**

## The Horse

Alexander the great is considered by many to have been the greatest military leader in history. He was a student of Aristotle, who was a student of Plato who had been a student of Socrates.

When he was ten years old he was brought to see a magnificent horse that many great horsemen had been unable to mount or train. The horse had been declared bad natured and untrainable – a lost cause. Alexander walked around the horse and noticed that it was frightened by its own shadow. He turned the horse toward the sun, hiding its shadow from it and calming it.

The horse's shadow had been there for all to see. Why was Alexander the only one to make the connection? Was it wisdom or lack of experience? Was it belief in himself or belief in the horse? Was it because he was not trying to prove any ideology or method, simply get to the heart of the matter? Why was Alexander the only one to think of looking where the horse was looking and to view the world from its point of view?

I don't know – I was not there. Maybe no one was. Maybe it was a foreshadowing of the life to come; maybe it is just a myth. There is one thing that is known: Alexander was famous for adopting the strengths of those he conquered.

## Reality As The Lever

The world is being ruled more and more by computers and process. The tools we make are allowing us to progress at an unprecedented rate. In the midst of this we are still monkeys with brains evolved to keep us safe from big cats and to find food amidst scarcity.

This brain allows us to make leaps into illogic to produce logic. Even in our mathematics it is necessary for us to lean into the imaginary. The square root of minus one is an imaginary number that cannot exist, first thought to be nothing but a curiosity but later discovered to be essential in the study and application of quantum theory, fluid dynamics, electromagnetism and other ideas central to our current understanding of reality.

Imaginary numbers are combined with real numbers to produce complex numbers. This offers us another dimension in geometry that allows us to access a kind of mathematical hyperspace.

## Humans As The Fulcrum

Humans can deal with change. We are difference engines. We can unconsciously deal with vast amounts of information that we don't need to connect directly with. We can approach great problems from an imaginary angle, and surface at their centre.

No matter what our conscious thinks is going on, it is always about what the unconscious perceives. Learning is consciously feeding the unconscious with instructions. The unconscious does not understand logic but it understands the deep structure of reality and strives to put us where we want to be in the chaotic, quantum soup that it protects us from.

Once we have moved on to the knowledge that the brain and human nature can change, and is built to change, we understand that activities can create changes in the structure of the brain and we are getting a clear glimpse of the sheer power of our ability to adapt.

## The Mathematics Of Change

Some of Newton's greatest gifts to us came to us through the mathematics of change: calculus. He discovered that integration (defining the area bounded by curves) and differentiation (the study of the rate at which quantities change) were the inverse of each other. The fundamental theorem of calculus declares that differentiation is the reverse of integration – its mirror.

Calculus opens a wormhole that allows monkeys, who have already come up with the complete works of Shakespeare, to hurl tin cans into space and out of the solar system. These tin cans contain improbable eyes that allow us to see further than any creature has ever seen.

## The Dog

> Imagine a sunny day in Ancient Greece. Flies are buzzing and there is the smell of heat in the air. Diogenes is relaxing in his favourite spot. He has had a pleasant morning of being seditious and crafty. Suddenly there is a voice. It declares that Alexander, the most powerful man in the world, has come to visit the famous philosopher.
>
> Diogenes rouses himself, looks up and sees the great king with his followers and says "I am Diogenes, the dog, and you are blocking my light."
>
> Alexander's followers must have held their breath. In the shocked silence Diogenes starts to scrabble around in a refuse heap.
>
> "What are you doing?" asks Alexander.

"Why, I am looking for the bones of your father. It appears that they cannot be distinguished from the bones of his slaves." replies Diogenes

Now there is a really deep silence from the onlookers. Some of them are even gleeful at the confrontation that must follow.

"Are you not afraid of me?" asks Alexander.

"Are you a good thing or a bad thing?" asks the man scrabbling in the dirt.

"A good thing." declares Alexander.

"Why should I be afraid of a good thing?" replies the old dog.

Diogenes decides to ask some questions of his own.

"What are your plans?" he asks.

"To conquer and rule all Greece." says Alexander

"What then?" asks Diogenes

"Conquer Asia Minor." comes the reply.

"And then?" asks Diogenes.

Alexander draws himself up looking toward the future and destiny.

"The world." he declares, stretching out his arms.

"The world." says Diogenes, shaking his head, "I am impressed."

Alexander nods. At last the old man seems to understand with whom he is dealing.

Diogenes pokes the refuse heap again.

"And when you rule the world, what will you do?" he asks.

Alexander thinks carefully.

"When I have done all this, as I will, I will rest and enjoy the rest of my life."

Diogenes repeats the answer.

"You plan to conquer all Greece, then Asia Minor and then the world so that you can relax and enjoy your life? Indeed. You do realise that you could save yourself and everyone else a lot of bother if you were to relax and enjoy life, here and now?"

It is said that Alexander sat in the refuse with him for the rest of the day and that as he left he declared that if he were not Alexander he would want to be Diogenes.

By his death Alexander had conquered the known world. By his death Diogenes had spent much of his life as a slave.

They both died in the same year, 323 BC, after very different lives

Alexander was 33 and Diogenes was 90 that year.

## Complex Numbers, Complex Systems – Reality Is Curved.

Human beings are complex systems. Societies are complex systems. Teams are complex systems. Being aware of the effect of actions on individuals and organisations and the cascade of effects is an essential part of life in our societies. Because this is about the measurement of change that is often not apparent or predictable from the initial conditions, we need to allow the flexibility to report feedback, whether it fits in with our expectations or not.

The best way to understand and manage consequences is to proceed in small steps, thereby keeping the feedback loop short. A small change in x over a small change in y is the rate of change or the differential.

## Boundaries

Induction separates the particulars from each other into groups of similarities. Deduction reasons from generalities into particulars.

## *Deduction*

If you start from a deductive initial condition, you must understand everything. This is the classical project management approach. You end up with too many objectives, all of which are possible end-games. You have to try to cover all the bases. It leads to the isolation and what is popularly called silo'ing of resources and effort. You isolate departments and effort and try to put it all together again at the end of the project. If you do not know everything – and you don't – all the king's horses and all the king's men will never put the product together again.

## *Induction*

If you start with an inductive initial condition you must assume nothing. This is the agile approach. You examine the pieces and try to generate the rules. You work with feedback. It can be time consuming and in the business environment it can create a perception of being too loose.

## *Combine induction and deduction at the boundaries*

The solution is to use deduction to create appropriate boundaries and to use induction to work within them. Deduction and induction are mirrors. As with order and chaos; simple and complex; integration and differentiation – they can cycle. Control passes from one to the other. Create the boundaries – the terms of reference – and work within them in iterations. Adapt and modify them on the basis of feedback.

The best way to understand consequences is to take small recursive steps. Neither a lender not a borrower be, neither an either nor an or be. Our strength is so connected to our ability to jump states that it would be a shame not to use it. If you have reached the bottom start chunking up[159], if you have

---

159 Moving to the more general or abstract.

reached the top start chunking down[160] again. The moving perspective must result in understanding. Understanding is the key that lets us out of the paradoxes that line our lives and our work.

## The Slave

> Diogenes took the position that conventional morality was all hypocrisy. He noticed that people were scandalised by the natural functions of their bodies and other people's bodies. At a time of gross injustice and real folly most people were very concerned and outraged by infractions of convention. Diogenes focussed on personal integrity and as a comment he went about the city with a lamp in broad daylight claiming to be looking for an honest man.
>
> He did not like Plato at all. While Plato was interested in the ideal of the truth, Diogenes was pragmatic and wanted to know what truth was when it came down from its stately pedestal. He declared, quite frankly, that he wanted to "stare truth in the arsehole".
>
> The term Cynic is derived from the Greek word for dog. He noticed that dogs live in the present and live without lies or hypocrisy. He noticed they live without apologies for their nature.
>
> When he was captured by pirates and sold as a slave, he was asked what he could do. He told them to warn whoever bought him that they were buying a master.
>
> Diogenes said that for the conduct of life we need right reason or a halter. The thing that frees the slave is reason. Even as a slave Diogenes was not a slave. His slavery was someone else's idea. His philosophy was one of adapting to circumstances. He adapted from being a self sufficient beggar to being a beloved teacher. He was referred to in the house as the "good genius". When his friends wanted to free him he said "Lions are not the slaves of those who feed them, but

---
160 Moving to the more specific or detailed.

> rather those who feed them are at the mercy of the lions: for fear is the mark of the slave, whereas wild beasts make men afraid of them.".
>
> This is a crucial knowledge. You can enslave yourself to anything: a job, a process, a lifestyle, an addiction, a fear of the world, a fear of other people or a fear of life. Reason is the choice that replaces fear. While we have an amazing capacity for fear we also have an amazing capacity for reason.

## The Truth About Happiness

It is no longer fashionable to scoff at personal happiness as a motivation, so it has become fashionable to portray happiness as complacent witlessness. It has always been fashionable to portray discomfort and suffering as the roots of creativity. This is a convenient lie. Happiness is linked to meaning and fulfilment. It is a changing target that abhors both complacency and witlessness.

Identity politics is just another way of trying to compare other people with your model. You can brand the wealthy as a self selecting club of selfish people who had to be completely unscrupulous to get in, or the poor as a load of wasters too lazy to succeed and spoiling it for the rest of us. These are just ways of lying to ourselves about our own motivations and what we think we should be doing.

Why do people want money? Why do they want power? Why do they want to keep up with the Joneses?

They hope it will make them happy, but it is pointless to build happiness on the slippery slope of where you stand in relation to some lifelong competition. If you do, happiness will always be just out of reach as you chase it. It perpetuates a philosophy of scarcity. There are some people who have decided that they can only be happy if they have more then everyone else. In doing this they lose sight of the intrinsic worth of what they have, because there will always be someone with more of something.

# The Truth About Progress

The balance point is not always the midpoint. The midpoint is usually the point at which the weaknesses of each approach cancel out the strengths of the other.

Committees, peer reviews and boards in science, medicine and business exist to make sure that the lunatic fringe is kept at bay.

If it were left to committees, peer reviews and boards then there would never be any change. They are a counterbalance that can often confuse the midpoint and the balance point.

Progress is often made by people who challenged the system and were a thorn in the side of the establishment (Newton, Einstein, Mozart, Bach Y Rita, Merzenich).

So how do we tell the nitwits from the genuine geniuses?

My advice is stop wasting your resources identifying nitwits and put it into discovering genius.

## *The Disney Pattern*

In NLP there is a powerful pattern called the Disney pattern by Robert Dilts[161]. It is based on the way the Disney company was known to work. It can be used as a planning tool.

You take the position of the extreme optimist. This is called the dreamer position. In this state you think of all the creative things that you could do. It is a fantasy state where there are no problems, constraints or costs.

When you have come up with all these great ideas, you take the position of the extreme pessimist. This is called the critic position. In this state you think of all the things that could go wrong and reasons why the ideas might not work. It is also a fantasy position where anything that could go wrong does go wrong.

---

161 "Strategies of Genius" by Robert Dilts

Traditionally you move backward and forward between these states to propose, challenge and answer. You can use different groups of people to represent the dreamers and the critics. In fact in most organisations in any meeting you go to this is played out. It can be a very combative experience as all proposals are tested to destruction. The battle usually goes to the most persuasive or domineering personalities and it is so much easier to break down than to build up. The critic position often attracts the most ambitious. It creates a tension between those motivated by authority and those motivated by achievement as defined by McClelland[162].

In the pattern there is also a realist position where you try to balance the dreamer and the critic. In the meeting room this tends to be someone who tries to balance the two and can often be someone motivated by affiliation as defined by McClelland.

This is not the way to balance; it is the way to détente or the midpoint where the weaknesses and strengths cancel each other out. When I have talked about filtering in this book I have pointed out that there are filters that add.

When we think of the Disney success we think of Walt Disney. Walt started the company with his younger brother Roy. Walt was a creative dreamer. Roy, however, was not a critic, but a balancer. He was the one who ensured the company was financially stable and the one who made sure there were ways to realise Walt's dreams. In his book "Building a Company: Roy O. Disney and the Creation of an Entertainment Empire" Bob Thomas tells us that without the partnership Walt would have ended up working for Walter Lanz and Roy would have been a bank manager.

The lesson from comparing the Disney Pattern to the reality of the Disney business is that you need a balancer. This is not someone who is trying to balance the dreamer and the critic. It is someone who is soundly on the side of the dreamer and

---

162 Discussed in "Triangulations and the power of 3" on page 148

who can take a meta position, learn from the critic what is required and make it happen.

## The Truth About Processes

Process has come to mean recipe. It involves writing down step by step instructions in the form of rules and constraints.

When businesses, self help gurus and get-rich-quick salesmen talk about process, they normally mean a recipe. It is a passive voice. It is dead, in the past, they did it this way and they were successful.

This is a complex equivalence. It is the process behind the process that you need to model. If you do, what you will find is someone who challenges the rules and asks why. You will find someone for whom success is a process of personal acuity leading to self actuation and the will to act whatever the personal inconvenience. You will find someone who understands that short term gain is not self interest and who can defer short term satisfaction to control long term gain. They can connect to other people and as a result they can connect to their future selves – the one we usually ignore who will be fat, unfit and unskilled. They are prepared to send gifts of skill, satisfaction and happiness into the future to that unknown self. They realise that they are a complex ongoing project.

Most complex things are never finished. That is their nature. They are never finished because the main attribute of complexity is its requirement to adapt to change.

Knowing that, let's bring process up to date. Process means doing it in the present.

Success is a process. Genius is a process. Intelligence is a process. Happiness is a process.

Process is the active voice. It is an active, ongoing, changing, evolving, process.

To process we need the ability to dissociate and to strive to apply logic. We need the ability to associate so that we can empathise and appreciate. In complex situations this means holding the real in one hand and the imaginary in the other and balancing them.

## The Truth About Techniques

There is a difference between picking and mixing techniques and understanding that techniques are the surface structure. It is the difference between a quick fix and a quick fix that works.

Good techniques uncover the real problems and filter them from the imaginary ones. They track the initial conditions. They uncover the why, the what and the how. They align them, continuously.

How you use a technique is more important than which particular technique you use as long as you are aware of the principles it is using. If a technique can be used to expose the deep structure and to enable congruence and maintain congruence with the natural flow, it is the right technique to use. If a technique allows you to apply collective intelligence to a problem rather than isolating and diminishing effort, it is the right technique to be using.

If, however, what you want to do is to introduce elements of a new method alongside your old method so that you can try it without risk, and you haven't examined the underlying principle, then you are likely only to introduce confusion.

## The Truth About Business

We can see that there are problems in the world caused in no little way by the advance of science and the tools with which it enables the greed of big business to run rampant. The problem is not with the scientific method but with the extrapolation of certainties drawn from inconclusive or incomplete data that we do not yet know is incomplete or inconclusive.

Business uses the discoveries of science that feeds the profit machine but ignores those that would calm its excesses. Many people are now starting to realise that the party is over and that business has to apply the science of consequences to itself.

Business can no longer try to isolate the science within its preferred frame from the science that is transforming our understanding of reality. The science of knowledge has moved way beyond anything that is currently being applied by managers in any company. We need to catch up and apply what we know to the places where we spend much of our conscious lives.

# Afterword

> *"When the way comes to an end, then change*
> *– having changed, you pass through."*
> **I Ching**

## Maintain Your Nerve

> "If you can keep your head when all about you
> Are losing theirs and blaming it on you,
> If you can trust yourself when all men doubt you,
> But make allowance for their doubting too;
> If you can wait and not be tired by waiting,
> Or being lied about, don't deal in lies,
> Or being hated, don't give way to hating,
> And yet don't look too good, nor talk too wise:"
>
> If – Rudyard Kipling

The time that I took to write this book has been difficult. I have been confronted with serious obstacles and problems outside my control. I really had to consider what I am writing here with a hefty dose of honesty. Sometimes the world pushes so hard that it feels like an unstoppable force and the tools of logic we build seem to be smoke in the face of a storm.

## Maintain Your Perspective

> "If you can dream – and not make dreams your master;
> If you can think – and not make thoughts your aim;
> If you can meet with Triumph and Disaster
> And treat those two impostors just the same;"
>
> If – Rudyard Kipling

Life is challenging and that is good. It has meant that I have been continuously faced with the need to use the tools I am talking about, even when I would have preferred to give up and rest.

## Maintain Your Vision

> If you can bear to hear the truth you've spoken
> Twisted by knaves to make a trap for fools,
> Or watch the things you gave your life to, broken,
> And stoop and build 'em up with worn-out tools:"
>
> If – Rudyard Kipling

I have had no shortage of inspiration and when I have been writing I have loved it. I would not swap one second of it. Whatever the outcome, the writing of this has been my life and I have spoken the truth as I have seen it and as it has whispered itself to me.

## Maintain Appreciation Of Chaos

> "If you can make one heap of all your winnings
> And risk it on one turn of pitch-and-toss,
> And lose, and start again at your beginnings
> And never breathe a word about your loss;"
>
> If – Rudyard Kipling

So what am I saying to you here? I think it is this, that there are times when we feel like we are carrying the world on our shoulders in the teeth of a hurricane. These are the times when we need to ask ourselves do we know what to do. Believe it or not, much of the hurricane happens either just in the future or just in the past. Most of us exist in the eye at the centre.

## Maintain Your Reality

> "If you can force your heart and nerve and sinew
> To serve your turn long after they are gone,
> And so hold on when there is nothing in you
> Except the Will which says to them: 'Hold on!'"
>
> If – Rudyard Kipling

Have you ever felt like this, that there is a gale blowing and you cannot make yourself heard over the noise? Have you thought of the last five minutes as the most thunder filled in your life? Does the act of just thinking that, make you realise the silence of now, sandwiched between the thunder of yesterday and the lightning of tomorrow?

## Maintain Your Self

> "If you can talk with crowds and keep your virtue,
> ' Or walk with Kings – nor lose the common touch,
> if neither foes nor loving friends can hurt you,
> If all men count with you, but none too much;"
>
> If – Rudyard Kipling

You can carry that silence with you. This is the centre that must hold. Rudyard Kipling's *If* is much quoted. I remember the first time I heard it and, as trite and old fashioned as it may sometimes seem, it still makes sense to me and makes me smile. He had something.

## Maintain Your Life

> "If you can fill the unforgiving minute
> With sixty seconds' worth of distance run,
> Yours is the Earth and everything that's in it,
> And – which is more – you'll be a Man, my son!"
>
> If – Rudyard Kipling

Philosophers over the years have striven to answer the questions of what we are, where we came from and why we are here. I am not so sure those are useful questions. They are rigged. They are loaded dice. In my opinion a better question might be: Now that I exist against all the odds, how can I fill each mercurial moment with conscious living, awareness and experience?

Answering that is everything there is to being. It is more than enough. Everything we are discovering about our consciousness tells us that it is through doing that we become ourselves and that this is how we finally become fully human.

# Appendix

## People Quoted At Chapter Beginnings

### Antoine de Saint-Exupery (1900–1944)

Writer, poet and aviator. Best known for his novella "The Little Prince" which as well as being a children's' book can also be interpreted as an astute analysis of human nature. St-Exupery was passionate about flying and flew mail planes over North Africa and the Andes. His books detail his own experiences and the inspiration he found in the sky and in the desert.

### Abraham Maslow (1908–1970)

Professor of psychology who founded humanistic psychology and created his hierarchy of needs to explain motivation. The hierarchy of needs goes from physiological (survival) to self actualisation. He studied Albert Einstein as one of the people he considered to be self actualised. He was one of the first psychologists to consider studying positive mental health rather than the ill or abnormal. His ideas concentrate on how people can access their own resources for growth and healing.

### Alfred North Whitehead (1861–1947)

Mathematician and philosopher who wrote the Principia Mathematica with Bertrand Russell in an attempt to provide the logical underpinnings of mathematics by showing that mathematics is reducible to formal logic.

### Aristotle (384–322 BC)

Philosopher and polymath who was a student of Plato and a teacher of Alexander the Great. He was the first to develop a formalised system for reasoning and is considered to be the founder of the field of logic.

## Bertrand Russell (1872–1970)

Logician, mathematician and philosopher generally recognised to be the founder of analytical philosophy. He wished to emphasize the scientific method in philosophy. Along with Kurt Gödel he is recognised as being one of the most important logicians of the 20$^{th}$ century. Together with Albert Einstein he launched the Russell-Einstein Manifesto calling for a curtailment of nuclear weapons. He defended neutral monism which rejects dualism. Unlike idealism (that there exists only the mental) and physicalism (that there exists only the physical) which also reject dualism, neutral monism holds that the there is a single type of substance and it can be viewed as mental or physical depending on context.

## Buddha (6$^h$ century BC)

Founder of Buddhism and who is said to have found enlightenment through meditation. The word Buddha means the one who is awake to reality.

## Diogenes of Sinope (412–323BC)

Philosopher with the best sense of humour in philosophy. He was the original cynic, which meant that he believed in self sufficiency, integrity self control, living by personal example and challenging conventional thinking particularly when it was hypocritical, conceited or corrupted. He constantly challenged the vanity, social climbing, self deception and artificiality of accepted conventions. His goal appeared to be to live a life of virtue free from possessions and to reject wealth, fame and power as unnecessary obstacles to happiness.

### Don McClean (1945 – )

American songwriter and singer. Wrote a number of iconic and metaphor rich songs including American Pie and Vincent.

### Dorothy Fields (1904–1974)

Broadway lyricist and librettist whose work was both sophisticated and down to earth and often gave cliché's a new slant.

### Edward de Bono (1933 – )

Physician, author and inventor. He is the originator of the concept of lateral thinking.

### Ernst Mach (1838–1916)

Philosopher and physicist who gave his name to the speed of sound and was admired by Einstein.

### Frank Herbert (1920–1986)

Author of the hugely successful Dune series of books. Herbert tackled subjects such as evolution, ecology and systems thinking in these books. He was interested in the work of Thomas Szasz who wrote a book about Karl Kraus called "Anti Freud" and who is famously critical of the moral and scientific foundations of psychiatry.

### *Frederic Chopin (1810–1849)*

Composer and virtuoso pianist who began composing at six years old and giving concerts at eight. He was a champion of absolute music which maintains that music is non representational and has no meaning other than its own aesthetics, formal structure and technical construction. Wagner and Hegel opposed this on the basis that they maintained that art had to have meaning. This is still a hotly debated topic and it seems to me that the definition of meaning is the contention. Music can be interpreted by the listener regardless of the conscious intent of the composer. This raises the question as to the nature of meaning[163].

### Gottfried Leibniz (1646–1716)

Philosopher and mathematician who is somewhat controversially credited with developing calculus independently of Newton and who invented the binary system used in modern computers. He introduced a granulated theory of matter he called Monads. He proposed that these monads were the smallest possible particles of force that were the basic living building blocks of the universe. He was grappling with the illusory nature of matter and he had a fairly workable theory of self similarity which Mandelbrot was to use in his development of fractal geometry.

### Gregory Bateson (1904–1980)

Anthropologist, social scientist, linguist, cyberneticist and thinker who was instrumental in extending and applying systems theory to behavioural science. He invented the idea of the double bind as a method to explain and address schizophrenia. He was very influential in the development of NLP. His father, William Bateson, coined the term genetics and popularised the work of Gregor Mendel, the friar who studied inheritance of traits in pea plants and who is posthumously credited with discovering the laws of genetic inheritance.

### Groucho Marx (1890–1977)

Comedian, film star and person who did not want membership in any club that would have somebody like him as a member. He insisted on singing for fifteen minutes

---

163 Look again at Rubenstein's comment on the music of Chopin on page 3.
   Perception is projection :)

during a visit to the New York Stock Exchange because he had lost eight hundred thousand dollars in the crash of 1929 and wanted to get his money's worth. The tickers ran blank for the duration.

## Hegel (1770–1831)

Georg Wilhelm Friederich Hegel, a philosopher who believed that we do not perceive the world directly but that our minds only access ideas about the world. He proposed that these ideas were in turn shaped by the ideas of other people. His ideas of thesis and antithesis are that every idea, thesis, contains an inherent contradiction, antithesis. The thesis and antithesis can be resolved into the synthesis which is a new idea containing elements of both the thesis and the antithesis. He was also concerned with the idea that humans are not only conscious but self conscious and that self consciousness is a social process in which we identify and take on the world view of another and in doing so becoming the object and the subject of consciousness – the consciousness being perceived and the consciousness doing the perceiving.

## I Ching (4$^{th}$ to 3$^{rd}$ century BC)

The book of changes is an ancient Chinese text. It contains a divining system which some people believe can enable you to use unconscious resources but is also interpreted as a philosophical work dealing with dynamic balance of opposites and the inevitability of change. Carl Jung was a fan and used it to develop his ideas of synchronicity which describes how some events can be related by meaning rather than causality.

## John Meynard Keynes (1883–1946)

Economist whose Keynesian economics fell out of favour in the 1970s after a period of general acceptance by most western industrialised countries since 1945. The economic collapse in 2007 has meant a renewed interest in his ideas since his main tenet is that the private sector should not be left to its own devices as it will be inclined to create inefficiencies. He said that insufficient demand causes unemployment and excessive demand causes inflation. He advocated government intervention in the form of spending and tax breaks to stimulate the economy in bad times and spending cuts and taxation to curb inflation in good times. At Versailles in 1918 he had tried to prevent the war compensation payments, being imposed on Germany by the treaty, from being so high, appealing to common sense and common humanity and predicting the second world war. Unfortunately for Germany and the world he was excluded from high level talks and his advice ignored. His book "The Economic Consequences of Peace" earned him the reputation of being anti establishment.

## Karl Kraus (1874–1936)

Writer, journalist, playwright, poet and satirist who considered language to be of great moral and aesthetic importance. He said that the language people used showed the real state of the world because he equated personal and political integrity with integrity of expression. He directed much of his satire at journalists and writers who used slovenly, pretentious or deceptive language. He detested rhetoric and considered Freud, psychoanalysts and psychoanalysis to be a disease posing as a cure and to be enemies of human dignity.

## Kurt Vonnegut (1922–2007)

Writer, humanist, artist and humorist who used science fiction to explore the world and the nature of existence. His book "Slaughterhouse 5, or The Children's Crusade: A Duty Dance with death" about time travelling aliens and his experiences as a prisoner of war during the fire bombing of Dresden is considered to be one of the greatest novels of the 20$^{th}$ century. His short story "Harrison Bergeron", which highlights the dangers of imposing equality, should be mandatory reading for all leaders, methodologists and managers.

### Leonardo Da Vinci (1452–1519)

All round genius and one of the most amazing minds ever to have graced the planet. Probably the greatest painter of all time and the very definition of scientist as he pursued and connected every field of knowledge. He fused art, science and engineering. He was fascinated by everything, curious about everything and he started each day with a list of things he would like to know. He observed the world around him and recorded his observations in amazingly beautiful journals. Not only was he ahead of his own time but we are still catching up with him.

### Mary Catherine Bateson (1939 – )

Writer, linguist and cultural anthropologist whose books deal with the impact of health and longevity on various aspects of society and with education, cross cultural communication, the environment and with her famous parents (Margaret Mead and Gregory Bateson).

### Noel Coward (1899–1973)

Painter, novelist, librettist, composer, singer, dancer, comedian, actor, stage producer, film director, cabaret artist, TV star and wit who was his own invention and has been a perennial influence on popular culture. His famous clipped style of speaking was developed, not as the result of an upper class background, but in response to his mother's deafness.

## Paul Bach Y Rita (1934–2006)

The first neuroscientist to seriously study neuroplasticity and introduce the idea of sensory substitution to treat people with neurological disorders.

## Philip K Howard (1948 – )

American lawyer and writer who comments on the effects of modern law and bureaucracy on human behaviour. He proposes reforms to the extremely litigative American legal system to create more freedom and to lessen the "corrosive fear" of litigation.

## Plato (424BC–348BC)

Philosopher and mathematician whose ideas laid the foundations for western philosophy and science.

## Pythagoras (around 570BC–650BC)

Philosopher, mathematician and founder of his own religion who is most famous for his theorem about right angle triangles. He is said to have set up a collaborative commune of like minded thinkers on the island of Croton. Bertrand Russell thought that Pythagoras should be considered the most influential of all western philosophers because of his influence on Plato and others.

## Rene Descartes (1596–1650)

Philosopher, mathematician, physiologist and writer who famously coined the phrase "I think therefore I am". He is considered to be the founder of modern philosophy. His attempts to describe the mind's response to external events by a flow of spiritual energy through nerves and his attempts to rationalise his religious beliefs with his scientific observation by locating the soul in the pineal gland led to an idea of dualism in western thought in which the consciousness is perceived as a spiritual non physical phenomenon. This idea of the mind being pure thought created the mind/body problem. He is also credited with inventing analytical geometry. He attempted to arrive at simple universal laws that govern all physical change and to devise a method of systematic doubt for arriving at the truth.

### Siegfried Sassoon (1886–1967)

Poet, writer and soldier decorated for bravery whose poetry describes the horror of war in the trenches of WWI. He was friends with Bertrand Russell, a pacifist, who encouraged him to send a letter to his commanding officer condemning the war as pointless and criticising those who, in Sassoon's view, were prolonging the war for their own ends of aggression and conquest. The letter was read out in parliament and Sassoon rather than being court martialled was treated for shell shock by Dr W.H.Rivers before returning to the front. Sassoon is recognised as one of the first writers brave enough to describe the reality of war as brutal, horrific and indefensible.

### Soren Kierkegaard (1813–1855)

Philosopher, theologian, psychologist and critic of Hegel and the German romantics. He is known as the father of existentialism. His work deals with religious belief and the concept of believing things without empirical proof. It also explores the relationship between objective and subjective experience.

### Tim Minchin (1975 – )

Distinctive musician and songwriter known for his atheism and his intelligent and sceptical style of comedic writing and performance.

## W Edwards Deming (1900–1993)

A statistician who increased productivity in the USA during the second world war and who is revered in Japan as a hero. His ideas were used to create the huge successes in Japanese manufacturing. His theories are the basis for Lean manufacturing.

## Winston Churchill (1874–1965)

Politician and statesman who led the United Kingdom during the second world war. He was also an artist, historian and writer. He is considered to be one of the great leaders of the twentieth century. He foresaw the rise of Hitler and the Nazis and during the war he built relationships with the American president Roosevelt, supported General de Gaul and maintained an alliance with the Soviet Union.

# Index

a posteriori............182, 197, 301
a priori...........77, 182, 195, 197p.
addressable memory..............98
agile..17, 111, 144, 170, 183, 210, 249, 259, 295p., 307, 316
Alexander the Great.....311, 328
algorithmic work...96, 239, 258, 260
alien abduction.......................67
Alzheimer's disease................28
American declaration of Independence........................63
American political system.....74
anthropomorphise................166
anvil of life..............................82
apartheid...............................278
Aquinas, Thomas..................148
Archimedes....104, 110, 157, 198
Aristotle....81, 146, 212, 311, 328
art......................................
    and learning...........................245
    and the "other" relationship................204
    of asking questons..............................144
    of balance..............................162
    of knowldege..........................173
    of science..............................150
    the arts.29, 119, 154, 163, 168, 182, 189, 200, 300
artificial intelligence......98, 141, 150, 175
association...........22, 69, 89, 100, 217p., 223, 271p.

associative arrays....................98
atomic distinctions...............132
auditory cortex.......................36
Australian desert..................309
autism.................35, 41, 188, 192
Babbage, Charles....................65
Bach Y Rita, Paul......19, 41, 254, 319, 335
bad habits..........................27, 31
Barry, Sue.............................146
Bateson, Gregory.....52, 87, 271, 331, 334
bereitschaftspotential ..........126
binary logic.............................98
Bohr, Neils............................108
boundaries......11, 17, 114p., 124, 131p., 152p., 155, 192, 235p., 240, 242pp., 255p., 272, 279, 292p., 295, 315p.
boundaries and constraints 192, 235
boundaries of reality............124
brain imaging.........................40
brain plasticity (see also neuroplasticity)....20, 31, 34p., 41, 54, 155
Broca, Paul.............................34
bundle theory .........................60
butterfly effect...............228, 309

*Note: p or pp indicates multiple occurrences on the same page.*

calculus............................313, 331

Calvin and Hobbes..................78

cargo cult................................166

chaos theory....18, 101, 229, 232, 235, 237, 241, 300

Chaplin, Charlie....................260

Chopin, Frederic....3pp., 8, 302, 330

cochlear implant...............27, 41

collaboration.....192, 233p., 304, 308

command and control...50, 232, 238

competitive plasticity...21p., 27, 39

complex equivalence.......30, 79, 83pp., 113, 137, 180, 208, 214, 321

complex numbers.........312, 315

connecting...4, 98, 120, 150, 156, 167, 170

conscious mind...8p., 49, 73, 75, 99, 102, 129, 139, 151, 280, 331p., 335

consciousness..............................
- *and artificial intelligence*....................141
- *and belief*..............................244
- *and business*..........................285
- *and creativity*.....................53, 309
- *and Freud*..............................32
- *and inspiration*......................151
- *and intuition*.........................101
- *and metaphors*.......................270
- *and mind/body*........................40
- *and neuroscience*......................19
- *and recursion*.........30, 126, 139, 169, 303
- *mechanisms of*.........................15
- *model of*................................6
- *nature of*...........................73, 190
- *self consciousness*.................175, 332

constraints.......17, 110, 114, 155, 192, 235p., 239, 241, 244, 250, 259, 283p., 292, 296, 301, 319, 321

control group............................64

corporate intrusion................263

corporate prosthetics............253

counterbalance. 73, 94, 142, 237, 242, 247, 260, 319

creative thinking.73, 205, 209p., 214

critical periods. 15, 34, 38p., 155

crucible of suffering 81p., 86, 89

culture......29, 166, 189, 233, 252, 281, 286, 303, 334

Curie, Marie........................157

curious water.........................133

Darwin,Charles..................185p.

Darwinian evolution............186

data compression. .22, 28, 301p.

Dawkins, Richard...................66

Deal or no deal...................110p.

deduction.....................182, 315p.

deep structure........18, 86p., 129, 146, 268p., 284, 300, 312, 322

deep structure of reality..18, 86, 312

deGrasse Tyson, Neil............296

delayed gratification...............90

demotivators.......................93p.

denominalisation.........199, 237, 288p., 304

Descartes, Rene......48, 162, 190, 311, 335
determinism............................23
deterministic universes........228
Deus ex machina...........180, 268
difference engine.....52, 65, 132, 312
Diogenes......61, 64, 313pp., 317, 329
Disney pattern...................319p.
Disney, Roy, O.......................320
Disney, Walt..........................320
dissociation.............23, 87, 271p.
distinctions.....11, 29, 41p., 64p., 86, 132, 163, 197, 202, 222, 257, 302, 308
distortion, deletion and generalisation...............140, 142
DNA..............................16, 184p.
Doidge, Norman.....................21
dynamite...............................194
ecology...........218, 246, 307, 330
Edison, Thomas.........63, 84, 150
effect of stress..........................49
effortless gain..........................84
Einstein, Albert...8, 101pp., 131, 138, 144, 156p., 169, 209p., 319, 328pp.
Einstein's brain......................156
emotional intelligence............18
empathy........63, 88, 188pp., 305
energy horses.......................108
epigenetics. 16, 177, 184pp., 189
Epimetheus..........................182

epistemology........................267
Escher, M.C..........................200
ethics..............................189, 307
evolution of knowledge. 121pp.
executive function.....24, 138pp.
extrinsic value.................96, 111
faces of versatility.................142
false belief task.....................190
Fastforward programme........42
feedback..................................
   *and balance*............................146
   *and boundaries*.....................316
   *and consciousness*................129
   *and data compression*............28
   *and empathy*.........................189
   *and induction/deduction*.....316
   *and knowledge*.............101, 136
   *and learning*...............9, 11, 27, 215, 217
   *and models*.....................87, 175
   *and pain*.........................46, 255
   *and projects*..........................164
   *and recursion*.......................169
   *and sensory substitution*................23, 27
   *feedback engineering*..........................255
   *feedback equation*............................230
   *feedback loop*......8, 37, 170, 197p., 215p., 227, 262, 281, 304, 315
   *feedback mechanism*...28, 46p., 169, 217, 295
   *feedback mirror*....................304
   *feedback principles*.............................217
   *sensitivity to feedback*................164, 302
   *tautological feedback*..........................169
FIFO......................................220
filtering....4, 22, 120, 130p., 133, 136p., 144, 151, 156, 163, 210, 214, 250, 320
financial markets..................235
Fleming, Alexander..............157
flow..8, 133, 212, 266, 269p., 304
fMRI........................................37

foresight and hindsight........182
fractal......................231, 282, 331
frame of reference. 131p., 207p., 210p.
Frankenstein............................77
free-loading..........................95p.
Freud, Sigmund 32, 75, 330, 333
frontier of knowledge. 15p., 296
fuzzy Logic..............98, 100, 276
Galileo...................................103
Galvani, Luigi Alyisio..........157
genetics....15pp., 21, 34, 56, 177, 184pp., 189, 331
genius 8, 150, 162, 166, 204, 212, 233, 302, 317, 319, 321, 334
giant hairball.................138, 252
Gödel, Kurt...104, 203, 252, 276, 281, 329
Graves levels..........................281
Graves, Clare.....................280p.
Hegel, Georg Wilhelm Friedrich......148, 175, 303, 330, 332, 336
Hemingway, Ernest................79
Herzberg, Frederic..................93
heuristic work.........96, 239, 260
hierarchy of needs.....280p., 328
hippocampus...........................50
Hobbes, Thomas......................78
Hubble telescope...................300
human advantage.................141
human factor.........................171
Hume, David...........................60
hypnotic trance.....................100

imaginary numbers..............312
incremental cycles.................168
induction....................182, 315p.
inertia........34, 37, 39, 130p., 139
information age.....116, 133, 210
initial conditions. 229p., 236pp., 244, 251, 259, 273, 283p., 315, 322
internal judiciary....................74
interpersonal communication ..............................................170
intersecting....111, 120, 146, 156
intrinsic value.......84, 90, 92pp., 96p., 111, 114, 116, 144, 165
intuition.9, 73, 98, 101, 138, 141, 209pp., 213
IQ.......................................17, 29
James, Tad.............................281
James, William.......................60
Kant, Immanuel...................148
kinaesthetic...........................222
king of the castle........264p., 306
Kipling, Rudyard............324pp.
knowledge workers.51, 97, 133, 261
ladder of life..........................161
Lamarck, Jean-Baptiste......185p.
Lamarckian evolution........185p.
lateral thinking 17, 210p., 276p., 330
left brain...........................73, 75
legislative function....75, 138pp.
levels of elevation..............297p.
levels of learning..................271

LIFO..................................220
Lincoln, Abraham....................63
litigation lottery.....................144
localisation..........................34, 38
Lock, John..............................61
logical wormholes..........208, 212
Lord Byron..............................65
Lucas, George........................233
Mach, Ernst.....124, 130pp., 146, 330
Mackenzie, Gordon...............252
Managing knowledge....60, 116
Mandela, Nelson....................278
Mandelbrot Set......................231
Mandelbrot, Benoit. .230p., 301, 331
Maslow Abraham......269, 280p., 310, 328
mathematics....6, 19, 101, 103p., 107, 110, 114, 182, 200, 203, 231, 276, 279, 281, 312p., 328
Matrix..........................................
    *film*.......................................................*271*
    *neo*........................................................*271*
    *red pill*................................................*268*
McClelland, David.........149, 320
McFarlane, Alan, Prof...........196
memes.......................................66
Merzenich, Michael........41, 319
mind body connection...35, 148
mind over matter.............48, 176
mirror in the mirror..............303
mirror neurons..............30, 187p.
Modern Times.......................260
money...83, 93, 95p., 113p., 272, 284pp., 318, 332
Morse code..................................9
motivators......................................
    *authority achievement affiliation*...148p., 320
    *motivation*.......50, 61, 93, 95p., 101, 129, 149, 165pp., 182, 190, 271, 277, 284p., 318, 328
    *motivator hygiene theory*......................*93*
motor cortex....................36, 187
multitasking............................26
Murray, Henry......................149
Muybridge, Eadweard.........109
nature or nurture....................16
negative reinforcement........216
neural state.............................70
neurogenesis.....................27, 30
neuroplasticity (see also brain plasticity).......15, 20p., 177, 335
neuroprogrammer..................20
neuroprogramming...........19pp.
neuroscience. .15, 18p., 31p., 35, 40p., 47, 49p., 157, 162, 170p., 177, 201, 203, 223, 254, 258, 264, 295p., 300
Newton, Sir Isaac........61, 109p., 228p., 248, 313, 319, 331
Newton's beach..................109p.
NLP..17, 70, 75, 89, 98, 111, 144, 149, 170, 183, 199, 222, 240, 245p., 259, 296, 308, 319, 331
noetics....................................211
nouvelle cuisine......................49
nun study................................28
Paine, Thomas........................63
Pandora...........................77, 182

paradox..........................................
   *creative paradox.................................162*
   *paradox of growth......................120, 159*
   *paradox of thought............................125*
   *plastic paradox............................38, 254*
   *Russell's Paradox..............................276*
   *uncertainty paradox...........................159*
parietal operculum...............156
peak experience..8, 150p., 269p.
pearl diver............................150
Penfield homunculus.............36
Penfield, Wilder....................36p.
perception is projection......165, 177
perceptual learning................29
perceptual positioning. 271, 281
phantom limb....................42, 46
phobias...............89, 244pp., 288
Pink, Dan................................95
Pixar.............................233, 265
Planck, Max..........................108
Plato....103, 146, 211p., 311, 317, 328, 335
positive reinforcement.........216
pre motor potential...............128
precedent law........................74
prejudice 20, 61p., 107, 136, 165, 169p., 226, 280
primary gain.........................183
principles of neuroplasticity. 21
productivity....95, 116, 197, 260, 262, 264, 301, 337
project management......61, 110, 162, 170, 277, 290, 305, 316
project manager....143, 162, 289
Prometheus..........................182

Prospero...............................278
protected self organisation
..........................................234pp.
psychiatry................66, 290, 330
psychosomatic illness...........177
quantum...........................
   *quantum mechanics..........................104*
   *quantum physics..................18, 109, 160*
   *quantum theory..................148, 300, 312*
Rainbow Nation....................278
Ramachandran, V.P, Prof 40, 42, 46pp., 87, 189
Rasmussen's syndrome..........54
recursion.....8p., 107, 168, 230p., 245, 289, 296, 303
requirements 118, 139, 207, 230, 236p., 249, 252, 261, 292p.
resilience of the brain..........54p.
resourceful state....................24
responses........................................
   *adversarial response...........................117*
   *drowning response............................117*
   *fight or flight response.......................240*
   *hoarding response.............................118*
   *requirements response.......................118*
reversed avatar.....................176
right brain........................73, 75
Rilke, Rayner Maria......153, 307
RIRO......................................21
risk compensation................193
risk homoeostasis.................193
rule of three.........................147
Russell, Bertrand....61, 103, 107, 203, 205, 276, 279, 281, 328p., 335p.
sacred cows..........................285

Sartre, Jean Paul....................306
scaling process......................258
Schrödinger's Cat...................228
secondary gain...27, 84pp., 179, 182p., 277, 284p., 288pp., 293
self actualisation...................328
self awareness.................175, 304
self organisation...............234pp.
self similar.....107, 229, 231, 241, 301, 331
semantics.....73, 179p., 201, 275, 277, 296
SEMSET....................................60
sensory substitution. 23, 41, 335
set theory........................103, 279
Shakespeare, William........267p., 278, 313
Shelley, Mary...........................77
shockproof crap detector.......79
skiing......................................247
slicing reality.................205, 207
Socrates..................................311
soften to harden....237, 247, 260
somatosensory.......................222
spacetime.....................208p., 282
step changes..........................168
strange loops..........200p., 203p., 209p., 289
Strawson, Galen......................60
streamline..............................264
stroke rehabilitation...............35
subconscious....................32, 75
submodalities........................222
superconsciousness..........8, 266

surface structure....18, 116, 129, 227, 229, 238, 267p., 284, 287, 322
suspension of disbelief.........243
synaesthesia..............41, 151, 222
synaptic connection................27
tea.....................................195pp.
terms of reference.........283, 316
tertiary gain...........................183
theory of constraints......17, 118, 257, 259, 332
theory of mind..178, 180p., 190, 192, 201, 305
Thomas, Michel.........................9
Thoreau, Henry, David........119
three dimensional sight.......146
three levels of focus..............284
throwing exceptions.............265
time's arrow..........................113
topographical.........................26
transderivational search...98pp.
transient epileptic amnesia....60
treehouse of truth.................242
triangulation............146pp., 297
unanchored metaphor..........124
unconscious mind....8p., 32, 52, 73pp., 99pp., 139p., 150p., 179, 211, 234, 240, 308, 312
unconscious skill...................234
values and beliefs......176, 267p.
vertical thinking.........210p., 276
visual cortex............................36
Voltaire................................61p.
Wales, Jimmy..........................95

waterfall.........144, 200, 277, 291
Whitehead, Alfred North......61, 103, 149, 281, 328
Wikipedia............................94p.
Wilson, Colin........................269
wire together, fire together....22
Woodsmalll, Wyatt...............281
wormhole 204, 208p., 211p., 313
Wright, Wilbur......111, 267, 333
zeitgeist, the.............................19
zero sum game........................62

# Bibliography

Bandler, Richard; Grinder, John – *The Structure of Magic* – Palo Alto (1976)

Barry, Susan; Oliver Sacks – *Fixing my Gaze: A Scientists Journey Into Seeing in Three Dimensions* – Basic Books (2009)

Bunting, Madelaine – *Willing slaves: how the overwork culture is ruling our lives* – HarperCollins (2004)

Dawkins, Richard – *The Selfish Gene* – Oxford University Press (1989)

De Bono, Edward – *Lateral Thinking; Creativity Step by Step* – Harper and Row (1970)

Descartes, Rene; (translated by John Cottingham) – *Meditations on First Philosophy* – Cambridge University Press (1996)

Dilts, Robert – *Strategies of Genius: Vol 1* – Meta Publications (1994)

Doidge, Norman – *The Brain That Changes Itself: Stories of Personal Triumph from the frontiers of brain science* – Viking Press (2007)

Ecenbarger, William – *Buckle Up Your Seatbelt and Behave* – Smithsonian Magazine (2009)

Graves, Clare – *Levels of Human Existence: An Open System Theory of Values,* – Journal of Humanistic Psychology (1970)

Herzberg, Frederick – *The Motivation to Work* – Wiley (1959)

Herzberg, Frederick – *Work and the Nature of Man* – World Pub Co. (1966)

Hofstadter, Douglas – *Gödel, Escher, Bach: An Eternal Golden Braid* – Basic Books (1979)

Iwerks, Leslie (Director) – *The Pixar Story* – Buena Vista Home Entertainment (2007)

Kohlberg, Jim (Director) – *The Music Never Stopped* – Roadside Attractions (Distributors) (2011)

Lanier, Jaron – *You are not a gadget: A Manifesto* – Alfred A Konpf (2010)

Levitin, Daniel, J – *This is your Brain on Music: Understanding a Human Obsession* – Atlantic (2006)

Lorenz, Edward – *Predictability: Does the Flap of a Butterfly's Wings in Brazil Set Off a Tornado in Texas?* – (1972)

MacKenzie, Gordon – *Orbiting the Giant Hairball: A Corporate Fool's Guide to Surviving with Grace* – Viking (1998)

Maslow, Abraham – *The Further Reaches of Human Nature* – Viking (1971)

Paine, Thomas – *The Rights of Man* – Dover Publications (2000)

Paine, Thomas – *The Age of Reason* – Dover Publications (2004)

Penfield, Wilder – *No Man Alone: A Neurosurgeon's Life* – Little, Brown (1977)

Penfield, Wilder; Rasmussen, Theodore – *The cerebral cortex of man* – Macmillan (1950)

Pink, Dan – *Drive: The Surprising Truth About What Motivates Us* – Cannongate Books (2009)

Pochmursky, Christina (Director) – *The Musical Brain* – National Geographic Channel (Worldwide distributors) (2009)

Premark, David & Woodruff, Guy – *Does a Chimpanzee have a Theory of Mind* – The Bahavioral and brain sciences. Vol. 1, no. 4 (1978)

Putnam, Hilary – *Reason, Truth, and History* – Cambridge University Press (1981)

Ramachandran, Vilayanur – *The Emerging Mind: The BBC Reith Lectures 2003* – Profile Books (2003)

Ramachandran, Vilayanur – *The Tell Tale Brain* – William Heinemann (2011)

Rogers, Kirsteen – *The Usborne Science Encyclopaedia* – Usborne (2009)

Sacks, Oliver – *The Man Who Mistook His Wife for a Hat and Other Clinical Tales* – Summit Books (1985)

Sacks, Oliver – *Musicophilia: Tales of Music and the Brain* – Alfred A Konpf/Picador (2007)

Sacks, Oliver – *The Last Hippie (An anthropologist on Mars : seven paradoxical tales)* – Picador (1995)

Schiller, Francis – *Paul Broca: Founder of French Anthropology, Explorer of the Brain* – University Of California Press (1979)

Shelley, Mary – *Frankenstein; or, The Modern Prometheus* – Lackington, Hughes, Harding, Mavor & Jones (1818)

Shenk, David – *The Genius in All of Us: Why Everything You've Been Told About Genetics, Talent and IQ Is Wrong* – Doubleday (2010)

Simmons, Dan; Chabris Christopher – *The Invisible Gorilla: how our intuitions deceive us* – HarperCollins (2010)

Snowdon, David – *Ageing with Grace: What the Nun Study Teaches Us About Leading Longer, Healthier, and More Meaningful Lives* – Bantam Books (2001)

Thomas, Bob – *Building a Company: Roy O.Disney and the Creation of an Entertainment Empire* – Hyperion (1998)

Wilson, Colin – *The Outsider* – Houghton Mifflin (1956)

Wilson, Colin – *Super Consciousness: The Quest for the Peak Experience* – Watkins Publishing (2009)

Witelson; Kigar; Harvey – *The Exceptional Brain of Albert Einstein* – The Lancet (1999)

Woodsmall, Wyatt; James, Tad – *Timeline Therapy and the basis of Personality* – Meta Publications (1989)

# Author Online

There is a website for this series of books at:

**www.TrousersOfReality.com**

where you will find the author's blog, resources, links and information related to subjects dealt with in the

*Trousers of Reality* Series.